Perspectives on Quantum Technologies - Modeling, Simulation, and Implementation

Edited by Eric Sakk

Published in London, United Kingdom

Perspectives on Quantum Technologies - Modeling, Simulation, and Implementation
http://dx.doi.org/10.5772/intechopen.1009696
Edited by Eric Sakk

Contributors
Akuonani Zakeyo Phiri, Anguang Hu, Babak Mohammadian, Chunyang Wang, Elson Rodger Katete, Evans Kumalo Mtambo, Francois-Igor Pris, Gilles Peslherbe, Hang Hu, Hsu Kiang Ooi, Jeolous Malamula Nyasulu, Mushtaq Rana Imran, Paul O'Hara, Zengxuan Zhao

First published in London, United Kingdom, 2026 by IntechOpen
IntechOpen is the global imprint of INTECHOPEN LIMITED, registered in England and Wales, registration number: 11086078, 167-169 Great Portland Street, London, W1W 5PF, United Kingdom

For EU product safety concerns: IN TECH d.o.o., Prolaz Marije Krucifikse Kozulić 3, 51000 Rijeka, Croatia, info@intechopen.com or visit our website at intechopen.com.

British Library Cataloguing-in-Publication Data
A catalogue record for this book is available from the British Library

Perspectives on Quantum Technologies - Modeling, Simulation, and Implementation
Edited by Eric Sakk
p. cm.
Print ISBN 978-1-83635-653-0
Online ISBN 978-1-83635-652-3
eBook (PDF) ISBN 978-1-83635-654-7

Meet the editor

Eric Sakk holds a Ph.D. from Cornell University in the field of Electrical and Computer Engineering. He is currently an Associate Professor of Computer Science at Morgan State University. Most recently, he has been involved in research on quantum computation and quantum communication. On the quantum computation side, he has published work on post-quantum cryptography, quantum implementations of Shor's algorithm, and quantum operators for elliptic curve applications. On the quantum communication side, he has published work on the performance of QKD systems based on quantum entanglement.

Contents

Preface

Quantum technologies are widely recognized for their influence across fields such as computation, communications, sensing, and energy systems. Technological advances in this ecosystem are rooted in a broad landscape spanning conceptual physics, mathematical modeling, computational simulation, and hardware implementation. *Perspectives on Quantum Technologies – Modeling, Simulation, and Implementation* reflects this layered structure, presenting contributions that illuminate various facets of the quantum technology realm.

The manuscripts in this work offer diverse perspectives. The chapters gathered here collectively demonstrate that quantum technologies evolve through the interaction of foundational insights, dynamical modeling, algorithmic innovation, and enabling hardware. Furthermore, the volume is organized in this progression.

The opening chapter, *Isotropic Bonding and Quantum Entanglement*, examines symmetry and entanglement as organizing principles in atomic and molecular systems. By exploring group-theoretical structures underlying entangled states, the chapter highlights the conceptual connections between quantum information and chemical bonding. This is followed by the *Perspective Chapter: Quantum-Like Modeling (QLM) and Its Philosophical Foundations*, which extends the reach of quantum formalism beyond physics, presenting contextual realism as a framework for understanding quantum-inspired modeling in complex systems and interdisciplinary settings. Together, these contributions establish a foundation that underscores the interpretive richness of quantum theory.

The next set of chapters focuses on dynamical processes and computational modeling. *Performance Analysis of Nonequilibrium Quantum Otto Engines in Relation to Quantum Coherence* investigates the role of coherence as a thermodynamic resource, demonstrating how nonequilibrium quantum effects influence energy conversion and system performance. Complementing this perspective, *Quantum Speedup in Molecular Integral Evaluation through Angular Momentum Coupling* introduces symmetry-aware computational methods that leverage angular momentum structure to improve efficiency in molecular simulations while suggesting connections to emerging quantum simulation platforms. These chapters highlight the importance of simulation and modeling as bridges between theoretical understanding and technological application.

The final chapter, *Superconducting Parametric Amplifiers: Resonator Design and Role in Qubit Readout*, turns attention to the physical realization of quantum technologies. By examining resonator architectures and parametric amplification mechanisms, this chapter emphasizes the critical role of quantum-limited measurement in scalable quantum information processing and the engineering considerations necessary for high-fidelity qubit readout.

Across these contributions, recurring themes emerge: the central roles of entanglement and coherence as quantum resources, the power of symmetry and mathematical structure in guiding modeling and simulation, and the necessity of precise measurement and control for reliable hardware implementation. The set of topics represented here reflects the scope of quantum technologies and the many pathways through which quantum ideas are translated into practical systems.

It is hoped that this volume will serve as both an introduction to current research directions and a catalyst for interdisciplinary dialogue. By presenting perspectives that range from models to simulation and, ultimately, to design, this collection seeks to foster collaborations between communities dedicated to the advancement of quantum science and technology.

The editor of this series would like to express sincere appreciation to the contributing authors for their thoughtful and innovative work, and to the reviewers and publishing team whose efforts supported the development of this volume. Their collective contributions have been essential in shaping a coherent and meaningful presentation of these diverse perspectives.

Eric Sakk
Department of Computer Science,
Morgan State University,
Baltimore, MD, USA

Section 1

Quantum Phenomena and Modeling

Chapter 1

Isotropic Bonding and Quantum Entanglement

Paul O'Hara

Abstract

With the advent of general relativity and quantum mechanics, physics has in many cases shifted the emphasis from dynamics to symmetry principles associated with different types of mathematical groups. We speak of the group structure of crystals: the group invariants of the equations of motion, the group classification of elementary particles, and also of the group homeomorphisms of fractals. The objective of this article is to show how the group theoretical structure of entanglement can be used to classify atomic and molecular structure. The article begins by establishing a working definition of entanglement and then uses it to derive the Pauli-exclusion principle by showing that Fermi-Dirac statistics can be considered as an entangled state invariant under the action of the $SL(n, C)$ group. A group classification scheme of entangled states is proposed to distinguish atomic and molecular bonds in chemistry. We conclude by discussing the possible use of interferometry in quantum computing.

Keywords: entanglement, group theory, Clifford algebra, atomic and molecular bonds, spin-statistics theorem, quantum computing

1. Introduction

Quantum entanglement is not an easy phenomenon to grasp. It involves both an understanding of probability and statistics, a distinction between pure and mixed states, the difference between indistinguishable and distinguishable states, an appreciation of L^2 functions, and a knowledge of group theory (which is often overlooked). Moreover, any serious discussion of the topic requires the use of a mathematical definition, and it is not always clear that it coincides fully with the physics understanding of the phenomenon. While most papers on the topic of entanglement dwell upon the notions of "locality" and "non-locality," entropy, Bell's inequality, and variations thereof [1], this paper is comparatively unique it that it focuses on the group theoretical properties of entangled states. Indeed, while recognizing "that the information in a composite system resides more in the correlations than in properties of individuals" [1], the article classifies these correlations using mathematical group theory to help us better understand the physical phenomenon of entanglement. In other articles, group theory, if mentioned, is only associated with a "covariant representation of spin and entanglement," [2], or by the "appropriate way to interpret the covariant entropy bound" [3]. Instead, in this paper, we classify entangled states using group theory and in particular show that there is a form of entanglement associated with the Bell states that help distinguish Fermi-Dirac and Bose-Einstein statistics, which can then be applied to the phenomenon of superconductivity and molecular bonding.

Before presenting a rigorous definition, we explore the meaning of entanglement from an intuitive point of view. Let us begin with a simple experiment. Imagine two people, one at the North Pole and the other at the South and each person is asked to look down and denote the direction with a vector. Both would probably write $|\downarrow\rangle$. However, if a neutral observer attempts to characterize the "looking down," s/he will notice that the two vectors are in opposite directions relative to one another and can be represented by either $|\downarrow\uparrow\rangle$ or $|\uparrow\downarrow\rangle$, depending on which convention is chosen to represent north and south. Moreover, if north and south are replaced by any two diametrically opposite positions on the sphere, the same dilemma arises. "Looking down" loses meaning in the sense what is "down" for one observer is "up" for the other and vice-versa. They are indistinguishable. In this case, classically speaking, it might be better to represent the two states by the vector $\frac{1}{2}(|\uparrow\downarrow\rangle + |\downarrow\uparrow\rangle)$ or $\frac{1}{2}(|\uparrow\downarrow\rangle - |\downarrow\uparrow\rangle)$, where $\frac{1}{2}$ indicates that each representation has equal weighting. In quantum mechanics, we replace the $\frac{1}{2}$ with $\frac{1}{\sqrt{2}}$ to reflect the Hilbert space structure of the wavefunctions (see below). The important thing is to be consistent throughout.

Either way, "up" and "down" have no absolute meaning, although the two vectors must point in opposite directions. The latter two representations can be considered more realistic in that they represent a superposition of two possibilities $|\downarrow\uparrow\rangle$ or $|\uparrow\downarrow\rangle$. There is no preference for one representation over another or of one direction over another. This becomes even more pronounced at the quantum level, where the two states are indistinguishable from one another, meaning that one cannot observe in principle "up" or "down" and "left" or "right" without disturbing the system. The act of observation usually forces us to impose a preferred reference frame, and doing so breaks the superposition. Also, the mathematical framework of quantum mechanics is formulated in terms of a Hilbert space, and in this case, as previously noted, it is preferable to write the superposed state as

$$|\psi\rangle = \frac{1}{\sqrt{2}}(|\uparrow\downarrow\rangle + |\downarrow\uparrow\rangle), \tag{1}$$

where $\frac{1}{\sqrt{2}}$ corresponds to the amplitude of the state vector and its square, $\frac{1}{2}$, to the respective probability weightings. We could also agree to write it as

$$\frac{1}{\sqrt{2}}(|\uparrow\downarrow\rangle - |\downarrow\uparrow\rangle). \tag{2}$$

The choice is ours. Indeed, someone could argue that it is better to write the joint superposed state as

$$\frac{1}{\sqrt{2}}(|\uparrow\uparrow\rangle + |\downarrow\downarrow\rangle), \tag{3}$$

where $\uparrow\uparrow$ corresponds to both of them "looking up" or $\downarrow\downarrow$ corresponds to both "looking down" depending on one's point of view. Either way, whether we use (1), (2), or (3) each represents an entangled state. As we will show later, all of these states are rotationally invariant in that they all represent "looking down" (or "looking up") from arbitrary positions on the globe provided they are diametrically opposite.

In many cases, including all rotationally invariant states, the entangled state is by definition a superposition of two indistinguishable states and if an experiment is carried out to distinguish one from the other, it means that the entanglement has been broken. Mathematically speaking, we could say that the wavefunction has collapsed. For example, in the case of (1), a measurement will cause $|\psi\rangle$ to transition into either the state $|\uparrow\downarrow\rangle$ or $|\downarrow\uparrow\rangle$. Usually in quantum mechanics, this is referred to as collapsing the wavefunction, and in effect, it means that we have transitioned from a superposed state to a separable or factored state. Completely factored states are non-entangled.

In general, it is common practice to distinguish a pure (including pure entangled) state from mixed and non-entangled states by means of the density matrix[1], which is defined as the outer product of the original states. With this in mind, we change notation and replace $|\uparrow\rangle$ and $|\downarrow\rangle$ by $|+\rangle$ and $|-\rangle$, respectively, with the understanding that $|+\rangle = \begin{pmatrix} 1 \\ 0 \end{pmatrix}$ and $|-\rangle = \begin{pmatrix} 0 \\ 1 \end{pmatrix}$. Now, consider the state

$$|\psi\rangle = p|+\rangle + \; q|-\rangle \tag{4}$$

with density function ρ_1 defined by

$$\rho_1 = \; |\psi\rangle\langle\psi| \tag{5}$$

$$= \begin{pmatrix} |p|^2 & pq^* \\ p^*q & |q|^2 \end{pmatrix}, \; |p|^2 + |q|^2 = 1 \tag{6}$$

Note that $\rho_1^2 = \rho_1$ in this case. If the wavefunction has collapsed, it means that the linear superposition has been broken and that it has been reduced to either the pure $|+\rangle$ or $|-\rangle$ with probabilities $|p|^2$ and $|q|^2$, respectively.

We can also define a density function that represents the probability ensemble of two states:

$$\rho_2 = |p|^2|+\rangle \; \langle+| + |q|^2|-\rangle\langle-| \tag{7}$$

$$= \begin{pmatrix} |p|^2 & 0 \\ 0 & |q|^2 \end{pmatrix}, \; |p|^2 + |q|^2 = 1 \tag{8}$$

It can be interpreted as an ensemble of two separable states, usually associated with identical particles. Sometimes, ρ_2 is referred to as a density matrix of mixed states. The transition from $\rho_1 \rightarrow \rho_2$ occurs experimentally when we assign probability weightings to the collapsed wave functions or more precisely to the probability distribution of mutually exclusive events.

The third point to note is that Eq. (4) could also have been written as

$$|\tilde{\psi}\rangle = p|+\;-\rangle + \; q|-+\rangle \tag{9}$$

[1] The density matrix seems to have been first introduced into quantum mechanics by von Neumann [4] and as noted by Tolman [5] the components ρ_{nm} of a density matrix "play a role somewhat similar to that of the density ρ in the classical mechanics"

corresponding to the fact that "up or down" at one pole can be simultaneously interpreted as "down or up" at the other pole. It is analogous to flipping a coin. If we observe that a "head" is up then simultaneously we know that a "tail" is down. One observation implicitly implies two pieces of information. This is also carried over into the density matrix for the superposed state (9), given by

$$\tilde{\rho}_1 = |\tilde{\psi}\rangle\langle\tilde{\psi}| \tag{10}$$

$$= (p|+\rangle|-\rangle + q|-\rangle|+\rangle)(p^*\langle+|\langle-| + q^*\langle-|\langle+|) \tag{11}$$

$$= \begin{pmatrix} 0 & 0 & 0 & 0 \\ 0 & |p|^2 & pq^* & 0 \\ 0 & p^*q & |q|^2 & 0 \\ 0 & 0 & 0 & 0 \end{pmatrix}, \quad |p|^2 + |q|^2 = 1 \tag{12}$$

It essentially carries the same information as ρ_1. Note that $\tilde{\rho}_1^2 = \tilde{\rho}_1$, $tr(\tilde{\rho}_1) = tr(\rho_1)$, and $det(\rho_1) = det(\tilde{\rho}_1)$ and there is a clear bijection between $\tilde{\rho}_1 \leftrightarrow \rho_1$. In other words, the previous comment that one observation yields two pieces of information carries over into density matrices.

Fourthly, this brings us to an important point regarding the principle of microcausality. Since an entangled (singlet) state simultaneously permits two pieces of information to be obtained from one measurement, the principle of microcausality is violated in this case. In other words, nonlocal events violate microcausality. It does not mean that a physical signal is transmitted faster (or indeed less) than the speed of light. It is the ontological observation that one measurement yields two pieces of information. In the case of a regular coin, if the head (H) is "up," then simultaneously we know the tail (T) is "down." The same applies to spin-singlet states. When the spin value of one particle is measured, then, simultaneously, the spin value of the other particle is known. This has serious implications for the spin-statistics theorem (cf Section 3.2) [6, 7].

2. Theoretical requirements and methods

2.1 A working definition of entanglement

Definition 1: Let $\mathcal{H}_1, \ldots \mathcal{H}_k$ be k (n-dimensional) Hilbert spaces such that $k \leq n$. Let

$$|\psi(x_1, x_2, \ldots, x_k)\rangle \in \mathcal{H}_1 \otimes \mathcal{H}_2 \ldots \otimes \mathcal{H}_k.$$

If there exists a set of states $|\psi\rangle_i \in \mathcal{H}_i$ such that

$$|\psi(x_1, x_2, \ldots x_k)\rangle = |\psi\rangle_1 \otimes |\psi\rangle_2 \ldots \otimes |\psi\rangle_k$$

then $|\psi(x_1, x_2, \ldots, x_k)\rangle$ is not entangled. Otherwise, it is said to be entangled.

This definition is in agreement with [8, 9]. The entangled states can be further subdivided into fully entangled states, which cannot be factored over $\mathcal{H}_1, \ldots \mathcal{H}_k$ and partially entangled states, which can be partially factored over $\mathcal{H}_1, \ldots \mathcal{H}_k$. For example, if $\mathcal{H}_1 = \mathcal{H}_2 = \mathcal{H}$ and $|+\rangle \in \mathcal{H}$ and $|-\rangle \in \mathcal{H}$ are orthogonal vectors then $|+\rangle|+\rangle$ and $|+\rangle|-\rangle$ are (pure) unentangled states, with respective density matrices given by

$$\begin{pmatrix} 1 & 0 \\ 0 & 0 \end{pmatrix} \text{ and } \begin{pmatrix} 0 & 0 \\ 0 & 1 \end{pmatrix}$$

and the corresponding mixed density matrix is given by ρ_2 (Eq. (8)). In contrast, the entangled state $|\psi\rangle = \frac{1}{\sqrt{2}}(|+\rangle|-\rangle - |-\rangle|+\rangle)$ is represented by the density matrix ρ_1.

As another example, consider the tensor product $\mathcal{H}_1 \otimes \mathcal{H}_2 \otimes \mathcal{H}_3$, then

1. $|\psi\rangle_1 \otimes |\psi\rangle_2 \otimes |\psi\rangle_3 + |\psi\rangle_2 \otimes |\psi\rangle_3 \otimes |\psi\rangle_1 + |\psi\rangle_3 \otimes |\psi\rangle_1 \otimes |\psi\rangle_2$ is entangled

2. $|\psi\rangle_1 \otimes (|\psi\rangle_2 \otimes |\psi\rangle_3 + |\psi\rangle_3 \otimes |\psi\rangle_2)$ is a partially entangled state.

3. $|\psi\rangle_1 \otimes |\psi\rangle_2 \otimes |\psi\rangle_3$ as a factored state is not entangled.

We have already noted in the introduction that many cases of entanglement can be associated with some degree of indistinguishability, although there are indistinguishable states that are not entangled (see triplet state defined below) and the mathematical definition does not per se refer to indistinguishability. In general, given $\mathcal{H} = \mathcal{H}_1 \otimes \cdots \otimes \mathcal{H}_k$, then $|\psi\rangle \in \mathcal{H}$ is entangled if it can be written as a non-separable orthonormal state

$$|\psi\rangle = \sum c_{i_1 \dots i_k} |\psi\rangle_{i_1} \otimes \cdots \otimes |\psi\rangle_{i_k}, \quad c_{i_1 \dots i_k} \neq 0 \ \text{ or } \neq 1$$

This is strictly a mathematical definition, and it remains an open question whether all such states are realizable in nature and whether such a definition truly captures the essence of entanglement as it occurs in physics. The previously defined set of states $|\psi\rangle_i$ can be combined to form a linear combination with a maximum of 2^k terms. This reduces to $k!$ terms, if $|\psi\rangle_i \neq |\psi\rangle_j$ for $i \neq j$. Moreover, if each $|c_i|$ is equal for all i, then the state $|\psi\rangle$ is indistinguishable, meaning that it is invariant under the action of a permutation group S_n, where n corresponds to the number of non-zero c_i terms. For example, if $k = 2$, we can write down the entangled state

$$|\psi\rangle = c_{11}|++\rangle_1 + c_{12}|+-\rangle + c_{21}|-+\rangle + c_{22}|--\rangle, \quad |c_{11}|^2 + |c_{12}|^2 + |c_{21}|^2 + |c_{22}|^2 = 1$$

which in the case of $|c_{11}|^2 = |c_{12}|^2 = |c_{21}|^2 = |c_{22}|^2 = \frac{1}{4}$ includes the triplet state

$$\frac{1}{2}(|+\rangle + |-\rangle) \otimes (|+\rangle + |-\rangle) \tag{13}$$

The triplet state serves as a good example of how mathematical and physical models may differ. Mathematically speaking, there are two ways to look at this. We view it as a tensor product of two states (or qubits), each of which is a linear superposition of $|+\rangle$ and $|-\rangle$, and write this product as

$$|\psi\rangle = \frac{1}{2}|++\rangle + \frac{1}{\sqrt{2}}\left(\frac{1}{\sqrt{2}}(|+-\rangle + |-+\rangle\right) + \frac{1}{2}|--\rangle \tag{14}$$

$$= \frac{1}{2}|1\rangle + \frac{1}{\sqrt{2}}|0\rangle + \frac{1}{2}|-1\rangle \tag{15}$$

with the understanding that $|1\rangle = \ |++\rangle, |0\rangle = \left(\frac{1}{\sqrt{2}}(|+-\rangle + |-+\rangle)\right)$, and $|-1\rangle = \ |--\rangle$ with respective probability weightings of $\{1/4, 1/2, 1/4\}$, or we view it as a superposition of three unit vectors $|-1\rangle, |0\rangle, |1\rangle$ defined over a three-dimensional vector space with respective probability weightings of $\{1/3, 1/3, 1/3\}$. Many physicists believe that the latter is the correct answer. The author of this article believes that a spin 1 state will have the probability decomposition $\{1/4, 1/2, 1/4\}$, given that the spin 1 state of atoms and mesons (composed of two quarks) arises from the presence of two spin 1/2-particles. The final answer will depend on the experiment.

In the case of a bivector composed of two orthogonal vectors, this reduces to a maximally entangled state of the form

$$|\psi\rangle = c_{12}|+\ -\rangle + c_{21}|-+\rangle, \ \ |c_{12}|^2 + |c_{21}|^2 = 1$$

which in turn reduces to a singlet state

$$\frac{1}{\sqrt{2}}(|+-\rangle - |-+\rangle)$$

when $c_{12} = \ -c_{21} = \frac{1}{\sqrt{2}}$.

It should be noted that each $c_{i_1 \dots i_k}$ can be either positive or negative and consequently by associating a probability measure with the L^2 property of $|\psi\rangle$, many of the usual properties of probability spaces are no longer valid [4], as is the case with the Chapman-Kolmogorov equations.

2.2 Bell states

The Bell states are defined as follows [9]:

$$|\psi\rangle_1 = \frac{1}{\sqrt{2}}(|+\rangle|+\rangle + |-\rangle|-\rangle) \tag{16}$$

$$|\psi\rangle_2 = \frac{1}{\sqrt{2}}(|+\rangle|-\rangle - |-\rangle|+\rangle) \tag{17}$$

$$|\psi\rangle_3 = \frac{1}{\sqrt{2}}(|+\rangle|+\rangle - |-\rangle|-\rangle) \tag{18}$$

$$|\psi\rangle_4 = \frac{1}{\sqrt{2}}(|+\rangle|-\rangle + |-\rangle|+\rangle) \tag{19}$$

First, note that the set $\{|\psi\rangle_1, |\psi\rangle_2\}$ is composed of conjugate spinors[2] as also is the set $\{|\psi\rangle_3, |\psi\rangle_4\}$. Secondly, all four Bell states subdivide into two sets of rotational invariant states. In other words, $(R,R)|\psi\rangle_1 = |\psi\rangle_1$ and $(R,R)|\psi\rangle_2 = |\psi\rangle_2$, while $(R,R^T)|\psi\rangle_3 = |\psi\rangle_3$ and $(R,R^T)|\psi\rangle_4 = |\psi\rangle_4$. In general, letting $|\psi\rangle = |\psi\rangle_1 + |\psi\rangle_2$ and $|\phi\rangle = |\psi\rangle_3 + |\psi\rangle_4$, it follows that

$$(R,R)|\psi\rangle = (R,R)|\psi\rangle_1 + (R,R)|\psi\rangle_2 = |\psi\rangle$$

[2] Some textbooks refer to $i|\psi\rangle_2$ and $i|\psi\rangle_4$ as the conjugate spinors. For the purpose of this article, we drop the $i = \sqrt{-1}$ for convenience because it substantially makes no difference.

and

$$(R, R^T)|\phi\rangle = (R, R^T)|\psi\rangle_3 + (R, R^T)|\psi\rangle_4 = |\phi\rangle \ .$$

Moreover, we note that the state

$$\frac{1}{\sqrt{2}}(|\psi\rangle_1 + |\psi\rangle_4) = \frac{1}{2}(|+\rangle|+\rangle + |-\rangle|-\rangle + |+\rangle|-\rangle + |-\rangle|+\rangle) \tag{20}$$

$$= \frac{1}{2}(|+\rangle + |-\rangle) \otimes (|+\rangle + |-\rangle) \tag{21}$$

is factorable over a subspace of the tensor product space $H_1 \otimes H_2$. It is not rotational invariant, and its probability density function is given by

$$\rho = \frac{1}{4}(|+\rangle + |-\rangle) \otimes (|+\rangle + |-\rangle)((\langle+| + \langle-|) \otimes (\langle+| + \langle-|) \tag{22}$$

$$= \frac{1}{4}\left[\begin{pmatrix} 1 & 1 \\ 1 & 1 \end{pmatrix} \otimes \begin{pmatrix} 1 & 1 \\ 1 & 1 \end{pmatrix}\right] \tag{23}$$

The reduced density functions are given by

$$\rho_1 = \rho_2 = \frac{1}{2}\begin{pmatrix} 1 & 1 \\ 1 & 1 \end{pmatrix} = \frac{1}{2}\begin{pmatrix} 1 & -1 \\ 1 & 1 \end{pmatrix}\begin{pmatrix} 1 & 0 \\ 0 & 0 \end{pmatrix}\begin{pmatrix} 1 & 1 \\ -1 & 1 \end{pmatrix} \tag{24}$$

Therefore, $Tr(\rho_1) = Tr(\rho_2) = 1$ and the von Neumann entropy [4] $Tr(\rho_1 \ln(\rho_2) = Tr(\rho_1 \ln \rho_2) = 0$. This means that the entropy of each subsystem is 0.

Again, if we consider the state

$$\frac{1}{\sqrt{2}}(|\psi\rangle_1 + i|\psi\rangle_2) = \frac{1}{2}(|+\rangle|+\rangle + |-\rangle|-\rangle + i|+\rangle|-\rangle - i|-\rangle|+\rangle) \tag{25}$$

$$= \frac{1}{2}(|+\rangle - i|-\rangle) \otimes (|+\rangle + i|-\rangle) \tag{26}$$

This state is rotationally invariant, the trace of each subsystem is 1, and their entropy is 0. However, if we consider the state $\frac{1}{\sqrt{2}}(|\psi\rangle_1 + |\psi\rangle_2$, it is not factorable but it is rotationally invariant and

$$\rho_1 = \rho_2 = \frac{1}{2}\begin{pmatrix} 1 & 0 \\ 0 & 1 \end{pmatrix} \tag{27}$$

which means that the entropy associated with ρ_1 and ρ_2 is equal and given by

$$-Tr(\rho_1 \ln \rho_1) = -Tr(\rho_2 \ln \rho_2) = \ln 2 \tag{28}$$

Consequently, we find that both of the states $\frac{1}{\sqrt{2}}(|\psi\rangle_1 + |\psi\rangle_4)$ and $\frac{1}{\sqrt{2}}(|\psi\rangle_1 + i|\psi\rangle_2)$ have entropy 0, but in contrast, the state $\frac{1}{\sqrt{2}}(|\psi\rangle_1 + i|\psi\rangle_2)$ is rotational invariant, while the state $\frac{1}{\sqrt{2}}(|\psi\rangle_1 + |\psi\rangle_4)$ is not. Also, $\frac{1}{\sqrt{2}}(|\psi\rangle_1 + |\psi\rangle_2)$ has a positive entropy, and it is also rotationally invariant. In other words, neither entropy alone nor rotational invariance alone are sufficient to distinguish between the different types of entangled and unentangled states. Something more is needed.

2.3 Isotropically spin correlated states

All four Bell states considered individually are not only rotationally invariant but also isotropically spin-correlated (ISC) [7, 10, 11].

Definition 2: More formally, n (quantum) particles are said to be isotropically spin-correlated (ISC), if a measurement made in an *arbitrary* direction on *one* of the particles allows us to predict with certainty the spin value of each of the other $n-1$ particles for the same direction.

For example, if two photons are in a singlet state, then no matter the orientation of the polarizer, the two photons will always be in equal and opposite polarized states. In other words, a singlet state means that the photons are polarized in every direction at once. There is no preferred direction. It is only by passing the individual photons through a polarizer that we impose a specific polarization. The same principle also holds for electrons in a spin-singlet state. There is no preferred direction. A measurement of spin value in an arbitrary direction will simultaneously yield an equal and opposite value for the other electron. The same logic also applies to all four Bell states. In the case of $|\psi\rangle_1$, the rotational invariance means that if we measure one value in an arbitrary direction in the plane then the other particle will have the same value if it were measured in the same direction. One might think of the state $|\psi\rangle_1$ as corresponding to parallel molecular bonds in chemistry and $|\psi\rangle_2$ corresponding to anti-parallel molecular bonds.

It is not immediately clear that ISC states must occur in pairs, which gives a privileged position to the four Bell states in that they may be considered the building blocks of all properly entangled systems. For example, if we were to consider by analogy that the singlet state corresponds to a right and left hand respectively, there is no reason we cannot draw three identical right hands and three identical left hands. We first draw one hand and then repeat the process ad infinitum to draw as many identical hands as we wish. We could say that the parameters that permit us to draw one right hand can be used over and over again to draw a second and then a third and so on. The same can be done with the left hand if needed. Using the same logic, one might expect that if we know all the parameters associated with determining the spin of an electron, then we could repeat the same procedure to produce n identical electrons. Einstein believed such parameters existed ([4], p. 209, [12, 13]) and that quantum mechanics as formulated was a mere cloak for ignorance and led "to spooky action at a distance." As it turns out, because of the implications associated with Bell's inequality [14], there are no hidden parameters and therefore spin states cannot be cloned. This leads to the following theorem.

Theorem 1: $|\psi_1\rangle$ and $|\psi\rangle_2$ are the only ISC states up to isomorphism.

A proof based on Bell's inequality is given in [7, 10, 11]. Indeed, the set $\{|\psi\rangle_1, |\psi\rangle_2\}$ as conjugate spinors represents parallel and anti-parallel pairing, respectively. The theory can also be formulated in terms of $\{|\psi\rangle_3, |\psi\rangle_4\}$. As noted in [11], "there is a duality principle at work which defines a $1 \leftrightarrow 1$ relation between the Hilbert spaces spanned by $\{|\psi\rangle_1, |\psi\rangle_2\}$ and $\{|\psi\rangle_3, |\psi\rangle_4\}$. For example, if we were to define ISC states in terms of invariance under $R(\theta) \otimes R^T(\theta)$, then by duality, we could apply Theorem 1 to $|\psi\rangle_3$ and $|\psi\rangle_4$." In this article, we will work primarily with $|\psi_1\rangle$ and $|\psi\rangle_2$.

As Definition 1 shows, mathematically speaking, there are many possible entangled states. However, in practice, when considering the world of physics, there is overwhelming evidence both from the theory of superconductors and molecular bonding theory that nature privileges pairwise ISC entangled states [15–32]. Moreover, as we have noted here and elsewhere, the Pauli exclusion principle can be proven by assuming the existence of indistinguishable ISC pairs. For this reason, we claim that ISC states are in many ways the building blocks of entanglement as it occurs in nature.

2.4 A Clifford algebra of entangled states

Although, entropy serves as a measure of entanglement [9], by itself it does not give a complete classification. It fails to distinguish between rotational and non-rotational states. Also, in the case of a bivariate system, all rotationally invariant states are constructed from the four ISC Bell states, which can be further subdivided into two disjoint Hilbert subspaces of $\mathcal{H}_1 \otimes \mathcal{H}_2$. Specifically, there exists a Hilbert space $\mathcal{H}_{12} \subset \mathcal{H}_1 \otimes \mathcal{H}_2$ with $X, Y \in \mathcal{H}$ such that $XY \in \mathcal{H}_{12}$ if and only if

$$XY \equiv \langle X, Y \rangle_{\otimes} + [X, Y]_{\otimes} \tag{29}$$

defines a Clifford Algebra[3] with respect to the basis $\{|\psi\rangle_1, |\psi\rangle_2\}$, with the understanding that

$$\langle X, Y \rangle_{\otimes} \equiv X \otimes Y + Y^{\perp} \otimes X^{\perp} \tag{30}$$

$$= (x_1 y_1 + x_2 y_2)(|+\rangle|+\rangle + |-\rangle|-\rangle) \tag{31}$$

defines an inner product and

$$[X, Y]_{\otimes} \equiv X \otimes Y - Y \otimes X \tag{32}$$

$$= (x_1 y_2 - x_2 y_1)(|+\rangle|-\rangle - |-\rangle|+\rangle) \tag{33}$$

defines an outer product. Note that the Clifford Algebra is rotational invariant with respect to $R \otimes R$. Indeed, in the case of a bivariate set of vectors, the above set of relationships can be identified with the complex numbers. To see this, consider $z_1, z_2 \in C$, then

$$\bar{z}_1 z_2 = (x_1 - i y_1)(x_2 + i y_2) = x_1 x_2 + y_1 y_2 + i(x_1 y_2 - x_2 y_1) \tag{34}$$

and note that $Re\,(\bar{z}_1 z_2) = \langle X, Y \rangle_{\otimes}$ and $Im(\bar{z}_1 z_2) = [X, Y]_{\otimes}$.

Similarly the dual basis $\{|\psi\rangle_3, |\psi\rangle_4\}$ can be used to define a conjugate Clifford algebra over a Hilbert space $\mathcal{H}_{34}$ given by

$$XY \equiv \langle X, Y \rangle_d + (X \wedge Y)_d \tag{35}$$

with respect to the basis $\{|\psi\rangle_3, |\psi\rangle_4\}$, with the understanding that $X = (x_1, ix_2)$, $X^c = (x_1, -ix_2)$ and $\tilde{X} = (ix_2, x_1)$ such that

$$\langle X, Y \rangle_d \equiv X \otimes Y - \tilde{Y} \otimes \tilde{X} \tag{36}$$

$$= (x_1 y_1 + x_2 y_2)(|+\rangle|+\rangle - |-\rangle|-\rangle)$$

defines an inner product and

---

[3] A description of the use of Clifford Algebra in physics can be found in [33]

$$X \wedge Y \equiv X^c \otimes Y - Y^c \otimes X \tag{37}$$

$$= (x_1 y_2 - x_2 y_1)(|+\rangle|-\rangle + |-\rangle|+\rangle) \tag{38}$$

an outer product. The conjugate Clifford Algebra is rotationally invariant under the action of $R \otimes R^T$.

Since the two Hilbert spaces $\mathcal{H}_{12}, \mathcal{H}_{34}$ are isomorphic to each other under conjugation, for the remaining of this article, we will work with the space $\mathcal{H}_{12}$. We note that the wedge product or the Lie product of the Clifford algebra $[X, Y]_{\otimes}$ is invariant under the action of $SL(2, C)$ group (see Theorem 2 below) and corresponds to a bivector. Entropy does not capture these properties. This raises interesting questions regarding the best classification of entangled states in general, especially in multivariate systems. For example, an entropy calculation applied to "The three forms of molecular oxygen" [24] will distinguish the different energy levels associated with the three forms but only the group theory will distinguish the three possibilities in terms of (1) the regular triplet state involving two π^* orbitals, (2) the singlet (ISC) state in the same π^* orbital and (3) the singlet state formed by paired electrons, one in each of the two different π^* orbitals.

3. Results

3.1 Molecular bonding

The existence of such paired ISC states raises an interesting question regarding the best classification of entantagled states especially in multivariate systems composed of molecules. In general chemistry a "bonding orbital" is produced when the energy of the molecular orbital is lower than the sum of the two individual orbitals from which it was composed. For example, when the electrons of two hydrogen atoms bind to form a molecular orbital bond of H_2 with spin state $|\psi\rangle_2$ (Eq. 17), it corresponds to the constructive interference of overlapping wavefunctions of the individual electrons, encapsulated by the singlet state with total spin 0. Consequently, the singlet has a lower energy than the sum of the two components taken separately. In contrast, an "antibonding orbital" occurs when the wave functions overlap out of phase (destructive interference in **Figure 1**) to produce a spin state $|\psi\rangle_1$. In this case, the bonding energy is higher than the sum of the individual energies associated with each separate state. Moreover, given that the spin states of $|\psi\rangle_1$ are parallel to each other, one can anticipate that this type of bond will have paramagnetic effects, while those of $|\psi\rangle_2$ will be diamagnetic. For the purpose of this article, we denote the molecular orbitals of the ISC states $|\psi\rangle_1$ and $|\psi\rangle_2$ by σ^* and σ, respectively. However, this differs from the usual use of the language in chemistry.

For example, in traditional molecular bonding theory, the molecule Li_2 will have the electron decomposition σ_{1s}, σ^*_{1s}, and σ_{2s}, whereas in this new formulation, it would be given by $2\sigma_{1s}$, σ_{2s} where $2\sigma_{1s}$ recognizes that there is σ_{1s} orbital associated with each atom, which one would also expect from symmetry, while σ_{2s} refers to the shared molecular orbital. Overall, the new notation indicates that there exists three spin-singlet states, each of which can be represented by $|\psi\rangle_2$.

As a second example, we consider the boron molecule B_2. Traditional molecular orbital theory lists the orbital structure as σ_{1s}, σ^*_{1s}, σ_{2s}, σ^*_{2s}, π_{2p_y}, and π_{2p_z} where π_{2p_y} and π_{2p_z} are referred to as unpaired electrons with parallel spins. In the new

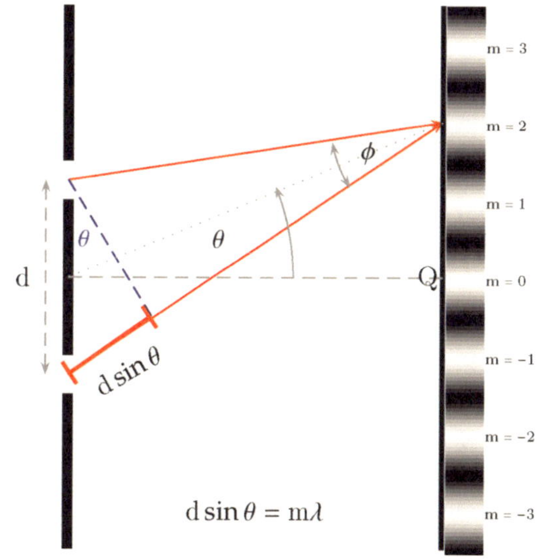

Figure 1.
Interference pattern.

representation, we write $2\sigma_{1s}$, $2\sigma_{2s}$, and σ^*_{2p} for the electron configuration. The presence of σ^*_{2p} also indicates the presence of paramagnetism, and the overall formulation respects the symmetry of B_2.

In general, the transition from two non-interacting electrons to interacting electrons by means of the chemical reaction can be considered as a phase transition from the triplet state (Eqs. 13, 14, 15) to the rotationally invariant Clifford algebra state given by Eq. (25), which in turn transitions to an equilibrium ISC state given by Eqs. (31) and (33), which form an orthonormal basis for the Clifford algebra. This transition from the Clifford state to one of the ISC states is mediated by the rule of "maximum multiplicity and minimal energy" [34]. In fact, a spin-singlet state represented by $|\psi\rangle_2$ might be expected to exhibit diamagnetic properties, while those in the parallel state represented by $|\psi\rangle_1$ might be expected to exhibit paramagnetism. The intensity of these effects will depend upon the physical closeness of the ISC states. As numerous experiments have demonstrated, the existence of ISC states are related to non-locality and can exist both over short-range and long-range distances. However, if the spin singlet is sufficiently close together, while inserted into a magnetic field one might expect (depending on their closeness) that the induced magnetic effect of each particle will be equal and opposite and result in them being repelled by each other, analogous to the repulsion between currents flowing in opposite directions in two parallel wires. In contrast, if the spin orbital is composed of two electrons in the parallel spin state, one might expect that when they are immersed in a magnetic field the induced magnetism will cause them to attract each other if they lie in the same plane perpendicular to the magnetic field. This is analogous to currents flowing in the same direction in two parallel wires [35]. Obviously, many other factors are involved including the orientations of all the different ISC states within the molecule, the electric repulsion among the electrons, temperature and pressure. In fact, the $|\psi\rangle_2$ state is invariant under the action of SL(2,C) regardless of orientation, while the parallel state is rotationally invariant over a plane, which gives a mathematical justification based on rotational invariance for the observation that "there is a nodal plane along the internuclear axis for all pi molecular orbitals" [34]. For this reason, when the σ^* orbitals (associated with the $|\psi\rangle_1$ state) are side-by-side, the paramagnetic effect should be enhanced and attractive.

The above examples, hopefully serve as a justification for replacing what is usually referred to in the literature as unpaired electrons with parallel spin, with the ISC state $|\psi\rangle_1$, which can be associated with a rotationally invariant pair of electrons in the same spin state. Moreover, the rotational invariance also enables us to predict the existence of spin-singlet states in the same orbitals (in the traditional sense of the word) and in different orbitals (again in the traditional sense of the word). The group theory puts no restrictions on the other quantum numbers. Consequently, two electrons can share the same quantum numbers (except for spin) meaning that they are in the same orbital or they can have different quantum numbers (meaning they are in different orbitals) but form a spin-singlet state. Indeed, this is what has been observed in "the three forms of molecular oxygen" [24]. To repeat, the group theory indicates the types of rotational invariance associated with chemical bonding. However, it imposes no other restrictions. In remains for the chemists to integrate this with the other laws of chemistry based on Hund's rules and the Pauli exclusion principle.

3.2 Fermi-Dirac statistics and entanglement

The Pauli exclusion principle is considered a key instrument in understanding chemical bonding and is often considered as a law in its own right. As it turns out, it can be derived by combining the notion of rotational invariance and indistinguishability. There are two approaches: The first is to show that the existence of n indistinguishable ISC states $|\psi\rangle_2$ is equivalent to the Pauli exclusion principle. A proof can be found in both [7, 10]. Indeed, as the indistinguishability condition is relaxed, different types of parastatistics emerge including identifying superconducting states with standing waves of ISC pairs. One class of superconductors can be associated with standing waves of multiple singlet states $|\psi\rangle_2$ and another with the rotationally invariant paired states $|\psi\rangle_1$. From this perspective superconducting Cooper pairs are a special case [11].

It has been shown that $|\psi\rangle_1$ and $|\psi\rangle_2$ are the only states that are ISC, and consequently, ISC states must occur in pairs. However, it might be asked if there are multivariate systems composed of indistinguishable particles that are rotationally invariant. As it turns out, an indistinguishable quantum state composed of ISC pairs obeys the Fermi-Dirac statistics. More precisely, it can be shown that a necessary and sufficient condition for Fermi-Dirac statistics is that the state be invariant under the action of $SL(n, C)$. Indeed, the Fermi-Dirac state is (uniquely) invariant under the action of the $SL(n, C)$ group, which a fortiori means that it is invariant for all subgroups of $SL(n, C)$, including the rotation group. It should be noted that, for two-dimensional subspaces $C^2 \subset C^n$, there are two rotationally invariant states and not one. This arises from the mathematical properties associated with Clifford algebras. In general, if **u** and **v** are two vectors in C^n, then we can define a Clifford product by

$$\mathbf{uv} = \mathbf{u}.\mathbf{v} + \mathbf{u}\wedge\mathbf{v}$$

The first term corresponds to an inner product which can be defined for any pair of vectors and such a product is always rotationally invariant. However, there is no inner product for three vectors **u**, **v**, and **w**, meaning that **u**.**v**.**w** is not defined. In contrast, outer (wedge) products are defined over n-dimensions and correspond geometrically to volume (area in 2-dim), which is preserved under rotations. This leads to the invariance of the Fermi-dirac state under SL(n,C) as proven in the following theorem [36]:

Theorem 2: Let $V = W_1 \otimes \cdots \otimes W_n$ be a vector space, with each W_i an n-dimensional subspace (up to isomorphism), $T = T_1 \otimes \cdots \otimes T_n$, with $T_i = T_j$ for each i and j, and T_i is a linear operator on W_i. If

$$v \equiv w_1 \wedge w_2 \wedge \cdots \wedge w_n \tag{39}$$

$$= \begin{pmatrix} w_{11} \\ \vdots \\ w_{n1} \end{pmatrix} \wedge \begin{pmatrix} w_{12} \\ \vdots \\ w_{n2} \end{pmatrix} \wedge \cdots \wedge \begin{pmatrix} w_{1n} \\ \vdots \\ w_{nn} \end{pmatrix} \tag{40}$$

then for $v \neq 0$

$$Tv = v \Leftrightarrow T \in \bigotimes_1^n SL(n, C).$$

This means the Fermi-Dirac statistics is invariant under the action of $SL(n, C)$.

Proof: For each W_i, we can associate an orthonormal basis $\{\mathbf{e}_1, \mathbf{e}_2 \ldots \mathbf{e}_n\}$. Therefore,

$$v = \begin{pmatrix} w_{11} \\ \vdots \\ w_{n1} \end{pmatrix} \wedge \begin{pmatrix} w_{12} \\ \vdots \\ w_{n2} \end{pmatrix} \wedge \cdots \wedge \begin{pmatrix} w_{n1} \\ \vdots \\ w_{nn} \end{pmatrix} \tag{41}$$

$$= |v| \mathbf{e}_1 \wedge \mathbf{e}_2 \wedge \cdots \wedge \mathbf{e}_n \tag{42}$$

where

$$|v| = \begin{vmatrix} w_{11} & w_{12} & \cdots & w_{1n} \\ w_{21} & w_{22} & \cdots & w_{2n} \\ \vdots & \vdots & \vdots & \vdots \\ w_{n1} & w_{n2} & \cdots & w_{nn} \end{vmatrix}$$

It follows from the linearity of T (that)

$$Tv = |v| T_1 \mathbf{e}_1 \wedge T_2 \mathbf{c}_2 \wedge \cdots \wedge T_n \mathbf{e}_n \tag{43}$$

$$= |v| \begin{pmatrix} t_{11} \\ \vdots \\ t_{n1} \end{pmatrix} \wedge \begin{pmatrix} t_{12} \\ \vdots \\ t_{n2} \end{pmatrix} \wedge \cdots \wedge \begin{pmatrix} t_{n1} \\ \vdots \\ t_{nn} \end{pmatrix} \tag{44}$$

$$= |v||T_1| \mathbf{e}_1 \wedge \mathbf{e}_2 \wedge \cdots \wedge \mathbf{e}_n, \qquad T_1 = T_2 = \cdots = T_n \tag{45}$$

But $v \neq 0$ implies $|v| \neq 0$, and therefore

$$Tv = v \Rightarrow |T_1| = 1 \text{ and } T_1 \in SL(n, C)$$

Conversely

$$T_1 \in SL(n, C) \Rightarrow Tv = v$$

The theorem has been proven. Moreover, v is unique as demonstrated in [36].

The observant reader might be wondering how this relates to Pauli's original paper on the spin-statistics theorem [6]. Pauli's proof as a point of departure assumes the validity of the principle of microcausality. Translated into operator language, this means that conjugate operators must commute beyond the light cone to preserve independence. However, in the case of ISC states, this breaks down. Operators and in particular Pauli's spin operators, indexed by different spacetime coordinates, are completely interchangeable because of the entanglement. In other words, the associated Clifford (Dirac) algebra may or may not commute beyond the light cone. In the case of ISC states the algebraic characteristics of the Pauli spin matrices applied to one particle of the bivariate state, instantaneously allows us to predict the spin characteristics of the second particle, which means that the information can be determined beyond the light cone.

Specifically, consider a two particle (bivariate) system $|\psi(\lambda_1)\rangle \otimes |\psi(\lambda_2)\rangle \in \mathcal{S}_1 \otimes \mathcal{S}_2$ where $\mathcal{S}_1 = \mathcal{L}^2(\mathcal{R}^3) \otimes H_1$ and $\mathcal{S}_2 = \mathcal{L}^2(\mathcal{R}^3) \otimes H_2$, respectively [7]. Note that each ket $|\psi(\lambda)\rangle \in \mathcal{S}$ can be written as $|\psi(q_1)\rangle \otimes s$, where s is a spinor. Also, let $\vec{S}_1 = (S_i(q_1), S_j(q_1), S_k(q_1))$ and $\vec{S}_2 = (S_i(q_2), S_j(q_2), S_k(q_2))$ be spin operators defined on the Hilbert spaces H_1 and H_2, respectively. If both particles are considered statistically independent of each other then individual measurements on each one are separated by a spacelike distance and will be represented, respectively, by operators of the form $S_i(q_1) \otimes I_2$ and $I_1 \otimes S_j(q_2)$. It follows, trivially that $[S_i(q_1) \otimes I_2, I_1 \otimes S_j(q_2)] = 0$, which means all measurements will commute beyond the light cone, as for example, in the case of the triplet state given by Eq. (13).

In contrast, the singlet state is ISC. In this case, it would be inappropriate to apply the operators for independent states. Instead, we let $|\psi(\lambda_1, \lambda_2\rangle \in \mathcal{S}_1 \otimes \mathcal{S}_2$ represent the spin-singlet state of two particles. Note that the perfect correlations between them allows us to put $H_1 = H_2 = H$ and to identify $\vec{S}_1$ and $\vec{S}_2$ as follows: Let $s_1(\theta)$ and $s_2(\theta)$ represent the spin states for particles 1 and 2, respectively, then for an arbitrary angle θ, there exists a unit vector $\vec{n}(\theta)$ such that $\vec{S}_1.\vec{n}(\theta)(s_1(\theta)) = \pm s_1(\theta)$ if and only if $\vec{S}_2.\vec{n}(\theta)(s_2(\theta)) = \mp s_2(\theta)$. This relationship allows us to identify $s_2(\theta)$ with the orthogonal complement $s_1^-(\theta)$ of $s_1(\theta)$ and to put $\vec{S}_1 = \vec{S}_2$.

This identification means that $\vec{S}_1.\vec{n}(\theta) = \vec{S}_2.\vec{n}(\theta)$ (note the same θ) and therefore $\left[\vec{S}_1.\vec{n}(\theta), \vec{S}_2.\vec{n}(\theta)\right] = 0$. However, because of the nature of entanglement, we are also free to consider $\vec{S}_1.\vec{n}(\theta_1)$ and $\vec{S}_2.\vec{n}(\theta_2)$ where $\theta_1 \neq \theta_2$, and in this case, $\left[\vec{S}_1.\vec{n}(\theta_1), \vec{S}_2.\vec{n}(\theta_2)\right] \neq 0$. In fact, $[S_i, S_j] = in\epsilon_{ijk}S_k$.

Simply put, microcausality does not apply to entangled states [7]. Entanglement permits both zero and non-zero commutators. It is an example, as noted in the introduction of one measurement yielding two pieces of information. Rather than asking if events are spacelike or timelike, we should first ask if they are correlated or uncorrelated and based on that answer then consider whether spacelike or timelike has any meaning.

Microcausality presupposes quantum locality. Entanglement as demonstrated by Bell's theorem [14] and verified experimentally by Aspect et al. [37] is a non-local phenomenon and consequently the principle of microcausality is violated by ISC states, where one measurement instantaneously yields two pieces of information.

Pauli makes no reference to entanglement is his paper. It was not an issue in his day. However, he does note that if the principle of microcausality fails then the statistical distinction between integral and half-integral spin also breaks down ([6], p. 721). The above theorem clearly implies that a superposition of n indistinguishable ISC states is equivalent to the Fermi-Dirac statistic, and that the Bose-Einstein statistic follows as a consequence of breaking the ISC condition. It is the ISC properties of spin or the lack thereof that determine the characteristics of quantum statistics. Spin value plays no role.

3.3 Group classification of entangled states

With the above theorems in place, we are in a position to classify entangled states using group theory. Specifically, let us distinguish between states comprising singlet states but with different degrees of reducibility with respect to $SL(n, C)$ group. Specifically,

$$c_1 SL(n_1, C) \oplus \cdots \oplus c_{n_m} SL(n_m, C) \Leftrightarrow \ |\psi\rangle = c_1 |\psi\rangle_1 \oplus \cdots \oplus c_m |\psi\rangle_m$$

where it is understood that $SL(n_i, C)|\psi\rangle_i = \ |\psi\rangle_i$ and c_i is the multiplicity. In other words, for any entangled state constructed from the Bell states (which are rotationally invariant), we can classify n-dimensional Bell entangled states according to the degree of indistinguishability that is permitted. If a state is invariant under the group $SL(n, C)$, then it will also be invariant under the group S_n. This follows because a Fermi-Dirac state is always invariant under the permutation group. For example, a set of six electrons could be configured to represent the six electrons of a carbon atom, or the five electrons of boron and one of hydrogen, or the four electrons of beryllium and two of helium, or the three electrons in two different lithium atoms. Each of these cases can be, respectively, classified as invariant under the action of the groups

$$SL(6, C), \quad SL(5, C) \oplus I, \quad SL(4, C) \oplus SL(2, C), \quad SL(3, C) \oplus SL(3, C),$$

which also serve as a specification of the degree of entanglement for each component. Note that the last case has multiplicity 2.

In the case of the molecules Li_2 with the orbital structure $2\sigma_{1s}$ and σ_{2s}, the group classification is given by $SL(2, \mathcal{C}) \otimes SL(2, \mathcal{C}) \otimes SL(2, \mathcal{C})$, where two independent lithium atoms have a group structure $SL(3, \mathcal{C}) \otimes SL(3, \mathcal{C})$. In the case of the boron molecule B_2, the bond structure is given by $2\sigma_{1s}, 2\sigma_{2s}, \sigma^*_{2p}$ which can be identified with the group structure $SL(2, \mathcal{C}) \otimes SL(2, \mathcal{C}) \otimes SL(2, \mathcal{C}) \otimes SL(2, \mathcal{C}) \otimes SO(2, \mathcal{C})$, where $SO(2, \mathcal{C})$ defines a two-dimensional rotational group.

3.4 Perfect correlations and ISC states

The preceding sections have focused on bivariate entanglement composed of ISC or rotationally invariant states. As we have noted, in the case of a bivariate system, there are essentially only two entangled states both of which are rotationally invariant and ISC. They can be combined to form a Clifford algebra which again maintains the rotationally invariant property. Nevertheless, it is important to point out that although ISC states are both rotationally invariant and perfectly correlated, they are two different concepts. ISC states in general refer to bivariate systems, while perfectly correlated states refer to multivariate systems involving three or more tensor products of single states. They are defined by Greenberger, Horne, Shimony, and Zeilinger [38] in their paper "Bell's theorem without inequalities" and are usually referred to as GHZ states. To further explore the meaning of entanglement within the context of multivariate states, we refer to their article (herein referred to by the acronym GHSZ). Another important study in this regard can be found in [7]. It might be worth noting that, in terms of entanglement in general, the Bell states serve as seed states for "entangled entanglement" and can be used to construct GHZ states through a process of iteration [39].

According to GHSZ [38], an entangled state is such that "it cannot be written in any way as a product of single-particle states"(p. 1132). This means that an entangled state cannot be written in the form

$$|\psi\rangle = \ |\psi_1\rangle|\psi_2\rangle \dots |\psi_n\rangle$$

This is is in agreement with Definition 1 of this article.

In general, if we are in the space $\mathcal{C}^3$, defined over the complex numbers, a threefold entangled state can be written as a linear combination of $3^3 = 27$ (complex) terms in the form

$$\sum \alpha_{ijk}|\psi_i\rangle|\psi_j\rangle|\psi_k\rangle$$

where each $i,\ j,\ k, \in \{1,2,3\}$. From a mathematical point of view, depending on the values of α_{ijk} there are an infinite number of three particle entangled states. In practise, we focus on certain specific states of interest to physics. For example, if $i \neq j \neq k$, then it will reduce to a state of only six terms, and if we further require that it be composed of indistinguishable terms then it will reduce to a Fermi-Dirac or Bose-Einstein state. It remains an open question whether every possible mathematically entangled state can be physically realized.

If $i,\ j,\ k$ are restricted to only two values (qubits), then the threefold entangled state can be written as a linear combination of $2^3 = 8$ terms, as is the case with the GHZ state defined by Eq. (G1) of GHSZ [38]. It is given by

$$\begin{aligned}\frac{1}{4}\Bigg[&\left(1 - ie^{i(\psi_1+\psi_2+\psi_3)}|d\rangle_1\right)|e\rangle_2|f\rangle_3 + \left(i - e^{i(\psi_1+\psi_2+\psi_3)}|d\rangle_1\right)|e\rangle_2|f'\rangle_3\\ &+\left(i - e^{i(\psi_1+\psi_2+\psi_3)}|d\rangle_1\right)|e'\rangle_2|f\rangle_3 + \left(-1 + ie^{i(\psi_1+\psi_2+\psi_3)}|d\rangle_1\right)|e'\rangle_2|f'\rangle_3\\ &+\left(i - e^{i(\psi_1+\psi_2+\psi_3)}|d'\rangle_1\right)|e\rangle_2|f\rangle_3 + \left(-1 + ie^{i(\psi_1+\psi_2+\psi_3)}|d'\rangle_1\right)|e\rangle_2|f'\rangle_3\\ &+\left(-1 + ie^{i(\psi_1+\psi_2+\psi_3)}|d'\rangle_1\right)|e'\rangle_2|f\rangle_3 + \left(-i + e^{i(\psi_1+\psi_2+\psi_3)}|d'\rangle_1\right)|e'\rangle_2|f'\rangle_3\Bigg]\end{aligned} \tag{46}$$

These states are particularly interesting in that they represent the evolution of the GHZ state given by

$$|\psi\rangle = \frac{1}{\sqrt{2}}\left[|a\rangle_1|b\rangle_2|c\rangle_3 + |a'\rangle_1|b'\rangle_2|c'\rangle_3\right. \tag{47}$$

where $|a\rangle_1 \to \frac{1}{\sqrt{2}}\left[|d\rangle_1 + i|d'\rangle_1\right]$ and $|a'\rangle_1 \to \frac{1}{\sqrt{2}}\left[|d'\rangle_1 + i|d\rangle_1\right]$. It should be apparent from this transformation that $|\psi\rangle$ is not rotationally invariant even when $\phi_1 + \phi_2 + \phi_3 = 0$, whereas when projected into a two-dimensional state with $\phi_1 + \phi_2 = n\pi$, it is rotationally invariant under (R, R^T), where R is a two-dimensional rotation matrix.

The general GHZ state for three particles presupposes that if we measure (observe) the states $|a\rangle$ and $|b\rangle$, then $|c\rangle$ would be determined by the other two. However, in the case of an ISC state, only one measurement would determine the other two. In other words, if we were to observe $|a\rangle$, then both $|b\rangle$ and $|c\rangle$ would be determined in the direction of measurement. However, as pointed out in Theorem 1 (Section 2.3), ISC states can only occur in pairs. The authors of the GHSZ paper, although they do not speak of ISC states per se, are aware that the Bell states differ from higher dimensional states. In their Appendix A [38], they clearly point out and prove the rotational invariance of the singlet state (Eq. (A3)), while in Appendix B [38] (although they do not develop it further), they imply that there is a difference between ISC states (my terminology) and "a rotationally invariant mixture of product states, which will not yield correlations as strong as [the singlet] $|\psi\rangle$ does." They then proceed to give an example.

Finally, we note that, in terms of GHZ states, rotational invariance is only applied to the two particle singlet state (paired qubits). There is no reference to other GHZ states being rotationally invariant, instead they refer to polarized states [38]. Apart from the bivariate state, the other GSZ states are not rotationally invariant, although they obey other group properties. For example, the GHZ states given by Eq. (48) and its evolved states are not invariant under rotations, but they are invariant under the action of the finite group of order eight $\mathcal{Z}_2 \times \mathcal{Z}_2 \times \mathcal{Z}_2$. In the case of entangled states that are NOT generated by the Bell states then the rotational invariance is lost, and each case should be classified according to its degree of polarization specified by the permutation group. For example, the perfectly correlated state

$$\frac{1}{\sqrt{2}}[|+\rangle_1|+\rangle_2|-\rangle_3|-\rangle_4 - |-\rangle_1|-\rangle_2|+\rangle_3|+\rangle_4] \tag{48}$$

is invariant (up to sign) under the group $\mathcal{Z}_2 \otimes \mathcal{Z}_2 \otimes \mathcal{Z}_2 \otimes \mathcal{Z}_2$.

3.5 Electron configuration and quantum computing

Quantum computing has an important role to play when probability based algorithms are employed to solve problems. Grover's algorithm [40], for example, when applied to a data base of size N can with a high probability find an item using $O(\sqrt{N})$ iterations rather than $O(N)$ iterations that would be required classically. It is a lot faster than a classical computer (although still of exponential order). Similarly, Shor's algorithm is probabilistic based [41]. In that regard, one would expect that the hardware for all probability-based algorithms within the immediate future will rely on entangled electron pairs to mediate the probability models. Currently, these ISC pairs are obtained from superconducting circuits or trapped ions. However, in what follows, it is proposed that paired ISC states are also at the core of interferometry and consequently, if this is true, offer another avenue to integrate hardware and software.

The advantage of using interferometry is that we can replace electrons with photons to achieve the ISC states, and this might substantially lower costs. In particular, if a beam of photons is passed through a Young's interferometer, the probability distribution can be analyzed in terms of the previously established coupling principle (Section 2.3). We prescind from the question of whether the individual photons passed through a single slit or both slits of the interferometer simultaneously (like a water drop) and instead focus on the quantum spin-state correlations as observed by the two apertures. These results will highlight the role of entanglement in interference phenomena, as well as clarify the role of conditional probability in physics.

3.5.1 A coupling principle

In Theorem 1, it has been shown that ISC states must occur in pairs and are effectively characterized by the Bell states (Eqs. (16)–(19)). For what follows, we focus on the Bell states $|\psi\rangle_1$ and $|\psi\rangle_2$ that together form a basis for a Clifford algebra. As previously noted, these entangled states have the characteristic that if a spin-measurement is made in an arbitrary direction on one of the states, then the state of the other particle is known with certainty in the same direction. Indeed, this article claims that the observed spinor (probability) correlations are the primary cause of the interference phenomena, and if these become uncorrelated (as often happens when we measure or try to detect a particle), then the interference phenomena disappears. Mathematically, it means that we are switching from applying the laws of conditional probability to the laws of independent probability.

Specifically, let the initial state

$$|\psi\rangle = \frac{1}{\sqrt{2}}(|+\rangle|+\rangle + |-\rangle|-\rangle) \tag{49}$$

represent the state as registered with respect to the normal at Q. (m = 0 in **Figure 1**). In other words, from the perspective of Q, we assume not only that the incoming particle has come with equal probability from either aperture but also that its spinor component along the normal to the screen at Q has been induced by the apertures working in tandem, as captured in the tensor product of Eq. (49). Next, shift the normal at Q to P (**Figure 2**). In this case, the pure initial state will transform under rotation operators $(\mathcal{R}(\alpha), \mathcal{R}(\beta))$ into the mixed quantum state

$$\begin{aligned}
&(\mathcal{R}(\alpha), \mathcal{R}(\beta))(|+\rangle|+\rangle + |-\rangle|-\rangle) = \\
&\begin{pmatrix} \cos(\alpha) & \sin(\alpha) \\ -\sin(\alpha) & \cos(\alpha) \end{pmatrix}\begin{pmatrix} 1 \\ 0 \end{pmatrix} \otimes \begin{pmatrix} \cos(\beta) & \sin(\beta) \\ -\sin(\beta) & \cos(\beta) \end{pmatrix}\begin{pmatrix} 1 \\ 0 \end{pmatrix} \\
+&\begin{pmatrix} \cos(\alpha) & \sin(\alpha) \\ -\sin(\alpha) & \cos(\alpha) \end{pmatrix}\begin{pmatrix} 0 \\ 1 \end{pmatrix} \otimes \begin{pmatrix} \cos(\beta) & \sin(\beta) \\ -\sin(\beta) & \cos(\beta) \end{pmatrix}\begin{pmatrix} 0 \\ 1 \end{pmatrix}
\end{aligned} \tag{50}$$

which on multiplying out and letting $\phi = \alpha - \beta$ gives

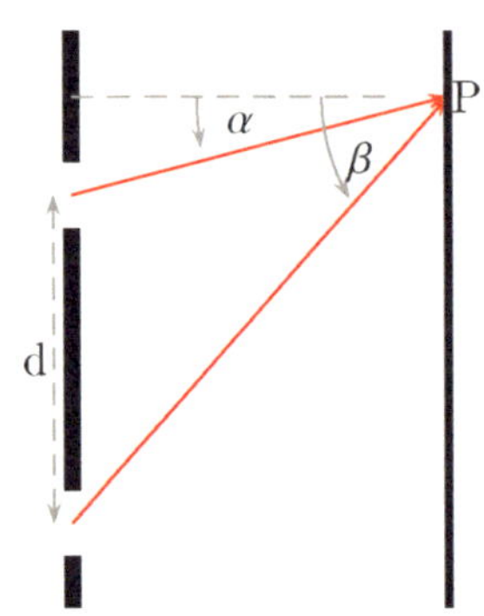

Figure 2.
Image seen at P.

$$\cos(\phi)\left[\begin{pmatrix}1\\0\end{pmatrix}\begin{pmatrix}1\\0\end{pmatrix}+\begin{pmatrix}0\\1\end{pmatrix}\begin{pmatrix}0\\1\end{pmatrix}\right] - \sin(\phi)\left[\begin{pmatrix}1\\0\end{pmatrix}\begin{pmatrix}0\\1\end{pmatrix}-\begin{pmatrix}0\\1\end{pmatrix}\begin{pmatrix}1\\0\end{pmatrix}\right]. \tag{51}$$

Note that ϕ and θ are related by the equation $\phi = \frac{2\pi d}{\lambda}\sin(\theta)$ (**Figure 1**) and that a binomial distribution, $b(N,p(\phi))$, can be associated with each choice of ϕ. Since, entangled pairs are at the core of quantum computers, the ability to produce entangled pairs using interferometry instead of superconducting circuits or trapped ions should encourage more research and investment in photonic quantum computers.

4. Conclusions

The article has pointed out that the rotational properties of quantum spin entanglement determine a coupling principle which states that isotropically spin-correlated states must occur in pairs. The no-cloning theorem follows as a special case. In turn, this pairing phenomenon can be used to derive and give a more intuitive understanding of the Pauli exclusion principle, serve as a unifying principle to explain conventional and unconventional superconductors, as well as bonding and anti-bonding interactions in chemistry and suggests a possible use in quantum computing. In addition, the article has discussed how group theory can add to our understanding of entanglement and molecular bonding. It has been pointed out that different types of entanglement can be associated with a group structure that leaves invariant the entangled state, and this structure can be associated with the degree of indistinguishability within the particle configuration. Conversely, if we know the group configuration of the particles, we can also specify the degree of indistinguishability among them. Group theory enables us to identify symmetries within the entangled state, although it does not distinguish which ISC state is more probable. In the bivariate case, in particular, the building blocks of entangled states are the two ISC states, which can be combined to form a Clifford algebra, with each component being, respectively, invariant under the action of the rotation group and the SL(2, C) group. Multivariate entangled states defined over $H_1 \otimes \ldots \otimes H_n$ also exist for $n \geq 3$ but with the exception of the Fermi-Dirac state, they are not rotationally invariant and are best classified by some type of finite group. Indeed, apart from the two ISC states, all other entangled states are polarized states with the entanglement applying to specific directions. The one exception is the Fermi-Dirac statistic which is SL(n, C) invariant and is composed of n ISC indistinguishable states.

Author details

Paul O'Hara
Istituto Universitario Sophia, Incisa e Figline Valdarno, FI, Italy

*Address all correspondence to: paul.ohara@sophiauniversity.org

References

[1] Brukner C, Zukowski M, Zeilinger A. The essence of entanglement. Fundamental Theories of Physics. 2001; **203**:117-138. DOI: 10.1007/978−3−030−77367−0_6

[2] Land M. Covariant representation of spin and Entaglement – A review and reformulation. Symmetry. 2024;**16**(11): 1465. DOI: 10.3390/sym16111465

[3] Adlam E. How are entanglement entropies related to entropy bounds. Philosophy of Physics. 2024;**2**(1):9

[4] von Neumann J. Mathematical Foundations of Quantum Mechanics. 1st ed. Princeton, USA: Princeton U; 1996 (12th printing)

[5] Tolman R. The Principles of Statistical Mechanics. NY: Dover Publ; 1979. p. 328

[6] Pauli W. The connection between spin and statistics. Physics Review. 1940; **58**:716-722. DOI: 10.1103/PhysRev.58.716

[7] O'Hara P. Entanglement, microcausality and Godel's theorem. Journal of Physics: Conference Series. 2023;**2482**:1-15. DOI: 10.1088/1742-6596/2482/1/012013

[8] Greenberger D, Horne M, Shimony A, Zeilinger A. Bell's theorem without inequalities. American Journal of Physics. 1990;**58**(12):1131-1143. DOI: 10.1119/1.16243

[9] Bennett C, DiVincenzo D, Smolin J, Wooters W. Mixed state entanglement and quantum error correction. Physical Review A. 1996;**54**:3825. DOI: 10.1103/PhysRevA.54.3824

[10] O'Hara P. Rotational invariance and the spin-statistics theorem. Foundation of Physics. 2003;**33**(9):1349-1368. DOI: 10.123/A:1025697412645

[11] O'Hara P. Spin entanglement - a unifying principle for superconductors and molecular bonding. Recent Progress in Materials. 2024;**6**:12. DOI: 10.21926/rpm.2402012

[12] Einstein A, Podolsky B, Rosen N. Can quantum-mechanical description of physical reality Be considered complete? Physics Review. 1935;**47**:777-780. DOI: 10.1103/PhysRev.47.777

[13] Wigner E. On hidden variables and quantum mechanical probabilities. American Journal of Physics. 1970;**38**: 1005. DOI: 10.1119/1.1976526

[14] Bell J. On the Einstein Podolsky Rosen paradox. Physics. 1964;**1**:195-200. DOI: 10.1103/PhysicsPhysiqueFizika.1.195

[15] Ueda K et al. Dominant nonlocal superconducting proximity effect due to electron-electron interaction in a ballistic double nanowire. Science Advances. 2019;**5**(10):1-6. DOI: 10.1126/sciadv.aaw2194

[16] Bardeen J. Electron-phonon interactions and superconductivity. In: Haken H, Wagner M, editors. Cooperative Phenomena. Heidelberg: Springer; 1973. p. 67. DOI: 10.1007/978-3-642-86003-4-6

[17] Bozovic I, Levy J. Pre-formed Cooper pairs in copper oxides and $LaAlO_3 - SrTiO_3$ heterostructures. Nature Physics. 2020;**16**:712-717. DOI: 10.1038s4157-020-0915-8

[18] Chaovsky P. Bose-Einstein condensation of rubidium atoms. JETP Letters. 2012;**95**(3):132-136. DOI: 10.1134/S0021364012030046

[19] Donley E, Claussen N, Thompson S, Wieman C. Atom-molecule coherence in

a Bose-Einstein condensate. Nature. 2002;**417**:529-533. DOI: 10.1038/417529a

[20] Georgescu J. 25 years of BEC. Nature Reviews Physics. 2020;**2**:396. DOI: 10.1038/s42254-020-0211-7

[21] Gezerlis A, Carleson J. Strongly paired fermions: Cold atoms and neutron matter. Physical Review C. 2008;**77**(3): 032801. DOI: 10.1103/PhysRevC.77.032801

[22] Huang H, Karan S, Padurariu C, et al. Universal scaling of tunable Yu-Shiba-Rusinov states across the quantum phase transition. Communications on Physics. 2023;**6**:214. DOI: 10.1038/s42005-023-01332-8

[23] Kovàcs A. An improved Calculation of Bose-Einstein CondensationTemperature. Modern Mathematical Physics. 2025;**1**(2): 1-6. DOI:10.3390/mmphys102000

[24] Laing M. The three forms of molecular oxygen. Journal of Chemical Education. 1989;**66**(6):453-455. DOI: 10.1021/ed066p453

[25] Madden R, Codling K. New autoionizing atomic energy levels in He, Ne, and Ar. Physical Review Letters. 1963;**10**:516. DOI: 10.1103/PhysRevLett.10.516

[26] Strontium unlocks the quantum secrets of superconductivity. SciTech Daily. NIST[Internet]. 2024. Available from: https://www.bing.com/ck/a?!&&p=60352f2d379fbddc718f425ff2b5d2439d29356d012ff0cbed7f00c0aac7a778JmltdHM9MTc2MjQ3MzYwMA&ptn=3&ver=2&hsh=4&fclid=31791e26-9904-6f38-3799-0f8898336e28&psq=Strontium+unlocks+the+quantum+secrets+of+superconductivity+ScviTech+daily&u=a1aHR0cHM6Ly9zY2l0ZWNoZGFpbHkuY29tL3N0cm9udGl1bS11bmxvY2tzLXRoZS1xdWFudHVtLXNlY3JldHMtb2Ytc3VwZXJjb25kdWN0aXZpdHkv

[27] Piestrup M, Puthoff H, Ebert P. Enhanced Correlated-Charge Field Emission. Palo Alto: Adelphi Technology Inc; 1998. DOI: 10.21236/ada337858

[28] Orlando T, Delin K. Foundations of Applied Superconductivity. Reading: Addison Wesley; 1991. 584 p.

[29] Ran S. Unboxing a New Spin-Triplet Superconductor. NIST[Internet]. 2019. Available from: www.nist.gov/blogs/taking-measure/unboxing-new-spin-triplet-superconductor

[30] Ran S et al. Nearly ferromagnetic spin-triplet superconductivity. Science. 2019;**365**:684-687. DOI: 10.1126/science.aav8645

[31] Karan S, Huang H, Padurariu C, et al. Superconducting quantum interference at the atomic scale. Nature Physics. 2022; **18**:893-898. DOI: 10.1038/s41567-022-01644-6

[32] Young D, Chu A, Song E, et al. Observing dynamical phases of BCS superconductors in a cavity QED simulator. Nature. 2024;**625**:679-684. DOI: 10.1038/s41586-023-06911-x

[33] Hestenes D, Sobczyk G. Clifford Algebra to Geometric Calculus. 1st ed. Dordrecht: Springer; 1984. 314 p.

[34] Witten K, Gailey K, Davis R. General Chemistry. 4th ed. Fort Worth: Saunders College, Publication; 1992

[35] Lorraine P, Carson D. Electromagnetic Fields and Waves. 2nd ed. San Francisco: Freeman Co; 1970. pp. 297-298

[36] O'Hara P. A generalized spin-statistics theorem. Journal of Physics Conference Series. 2017;**845**:012030. DOI: 10.1088/1742–6596/845/1/012030

[37] Aspect A, Dalibard J, Roger G. Experimental test of Bell's inequalities using time varying analyzers. Physical Review Letters. 1982;**49**:1804-1807

[38] Greenberger D, Horne M, Zeilinger A. Going beyond Bell's theorem. In: Kafatos, Menos, editors. Bell's Theorem, Quantum Theory, and Conceptions of the Universe. Heidelberg: Springer; 1989. p. 69-72

[39] Uchida G, Bertlmann R, Hiesmayr B. Entangled entanglement; a construction procedure. Physics Letters A. 2015; **379**(42):2698-2703. DOI: 10.1016/j.physleta.2015.07.045

[40] Grover K. A fast quantum mechanical algorithm for database search. In: Proceedings of the Twenty-Eighth Annual ACM Symposium on Theory of Computing - STOC '96. Philadelphia, Pennsylvania, USA: Association for Computing Machinery. 1996. p. 212-219. DOI: 10.1145/237814.237866. ISBN 978-0-89791-785-8. S2CID 207198067

[41] Shor P. Polynomial-time algorithms for prime factorization and discrete logarithms on a quantum computer. SIAM Journal on Computing. 1997;**26**(5): 1484-1509. DOI: 10.1137/S0097539795293172. S2CID 2337707

Chapter 2

Perspective Chapter: Quantum-Like Modeling (QLM) and Its Philosophical Foundations

Francois-Igor Pris

Abstract

We argue that the formalism of quantum mechanics can be fruitfully applied to the modeling of phenomena and systems beyond the domain of physics—particularly in the social sciences and humanities. As is well known, this is indeed the case; Andrei Khrennikov's quantum-like modeling (QLM) offers a prominent example. In this chapter, we propose a philosophical grounding for QLM within the framework of contextual realism (CR). When applied to quantum physics, this framework takes the form of contextual quantum realism (CQR). At the heart of CR lies the idea that ontology is sensitive to context, and that contextuality is inseparable from normativity. We offer a brief overview of CR and focus on its relevance to the foundations and potential of QLM.

Keywords: quantum mechanics, social phenomena, humanities, quantum-like modeling, contextual realism

1. Introduction: Quantum-Like Modeling (QLM)

Khrennikov's QLM applies the formalism of quantum mechanics (QM) and quantum field theory (QFT) to the modeling of non-physical phenomena and systems—biological, mental, social, economic, ecological, political, and others. These systems are generally *open* [1], meaning they interact with their environments or other systems; accordingly, the formalism of open quantum systems is employed.

Importantly, this approach does *not* involve reducing macroscopic non-physical phenomena to microphysical quantum ones, nor does it explain them on the basis of a fundamental quantum ontology of atoms and elementary particles. While such localized reductions may be viable in disciplines such as biophysics, neurophysiology, or cognitive science, the reduction of social, economic, or political systems to quantum microphysics is generally unwarranted.

In broader terms, the social sciences and humanities are not reducible to the natural sciences. Biological, social, economic, political, and media domains constitute distinct realities—each autonomous from physical reality and from one another. The rejection of metaphysical reductionism (which we are cautious to call anti-reductionism, so as to avoid a metaphysics of the opposite kind) nonetheless allows for local, context-sensitive reductions where appropriate. This stance does *not* entail a rejection of a unified naturalistic—or broadly scientific—methodological approach.

In this sense, the framework remains *monistic* in its understanding of the natural sciences, social sciences, and humanities, while maintaining compatibility with *pluralism*. Cognitive sciences, for example, have made a significant contribution to the unified picture of scientific knowledge.

QLM is viewed by its proponents as a universal monistic approach to a wide variety of areas of reality.[1] Our own philosophical position—contextual (quantum) realism (CR/CQR)—similarly asserts a claim to universality [2–6]. We will argue that CR/CQR provides the philosophical foundation for QLM.[2]

2. Contextual realism (CR)

CR takes as its first postulate (1) the assertion of the primacy (fundamentality) and unambiguity of the concept of reality. Reality is just what it is. (This is an analytic proposition expressing the grammar of the concept of reality). In other words, the real is real; it is what is. As Jocelyn Benoist says, "Reality *simpliciter* is just *what norms apply to.* As such, it is what it is, whether a norm is applied to it or not" ([8], p. 181). The given—the visible, that which simply appears and may therefore be illusory—is secondary. Appearance aligns with reality, or proves to be reality, if and only if the norm of reality is satisfied. This norm is ideal: It does not belong to reality itself but functions as a kind of "movement" within it—a transition from one sensible configuration to another.

Thus, the second postulate of CR can be stated as follows: (2) The categorical dualism of the ideal—particularly the normative—and the real. Norms, rules, concepts, meanings, attitudes, and theories belong to the realm of the ideal. Yet they are rooted in experience, practice, and reality—that is, they possess real conditions of applicability. This reflects a realist stance on norms and the ideal more generally.

We refer to such rules (norms, concepts) as *Wittgensteinian rules* (or W-rules), drawing inspiration from the later philosophy of Wittgenstein. By applying W-rules, we measure reality—reality being that which is measured by norms—within language games (LG), or phenomena, governed by those very rules.

When the norm is satisfied—that is, when the language game functions correctly—what appears is real [2–6].

CR affirms the existence of an infinite plurality of kinds of reality while maintaining that all such kinds are real in the same sense. Psychological, social, economic, and media realities are neither more nor less real than physical or biological reality. Differences in the *nature* of reality should not be misconstrued as differences in the *definition* of reality—which remains unambiguous—or as differences in its *degree*.

---

[1] QBism also lays claim to this role. Earlier, we showed that it represents an anti-realist position, which, however, can be revised within the framework of our CQR.

[2] S. Patra and S. Yeddanapudi's approach [7] likewise aspires to serve as a philosophical foundation for QLM (see our critique of this approach below).

Thus, CR serves as common ground across the social sciences, the humanities, and the physical sciences—that is, across diverse domains of phenomena and systems. In particular, the question of whether it is possible and meaningful to transfer concepts, rules, and norms from physics beyond its disciplinary boundaries, and to radically generalize them can be addressed as follows: Such transfers (or generalizations) are legitimate only if there exist concrete conditions for the use of the generalized concepts, rules, and norms—or, in Wittgenstein's terms, if there are actual language games in which they function. In this respect, CR may be understood as a form of normative pragmatism and broad naturalism.

The initial metaphorical transfer of concepts across domains enables us to gesture toward phenomena not yet named, to establish new norms and meanings, and to articulate emergent ontologies.[3] The "legalization"—so to speak—of these expanded concepts occurs at a subsequent stage, when their actual use becomes embedded in a stable practice. This holds for the importation of physical concepts into the socio-humanitarian sphere, but it is equally true within physics itself: The extension of classical mechanical concepts gave rise to quantum mechanics and the formation of novel quantum concepts. The generalization of the W-rule exemplifies this dynamic, as it leads to fundamentally new applications and thus evokes the Wittgensteinian problem of rule-following—not merely as a question of applying a rule, but as the challenge of transitioning from one rule to another.

Ontology, understood as a determinate real existence in which appearance coincides with reality, emerges through the measurement of reality by contextually applied rules (norms)—rules whose application, by definition, constitutes judgment, and judgment is inherently contextual. Accordingly, both ontology and epistemology are context-sensitive. Norms—that is, the ideal—precede cognition and ontology (in the ontic sense), just as epistemology precedes ontology (again, in the ontic sense). That being said, norms, epistemology, and (re)cognizable ontology are all secondary to reality as such.

Normativity (the ought), conceptual distinctions, and contextuality (sensitivity to conditions) always accompany one another. Likewise, where there is no contextuality (normativity or conceptuality), one cannot speak—or even think—of reality, of anything real. As Benoist writes, "Context sensitivity is a fundamental epistemological and ontological property of reality" ([10], p. 53).

At the heart of CR lies the claim that contextuality, far from implying relativism, is to be conceived as inseparable from normativity and conceptuality. This demands a departure from the philosophical paradigms of both modernity and postmodernity, while drawing critically on the resources of continental and analytic traditions—each of which, in its own way, remains entangled with forms of anti-realism. CR aims to

[3] A case in point is the introduction into cognitive science of the poorly defined concept of information, conceived not merely as knowledge but as a new kind of reality. As Daniel Andler notes, cognitive science "came into being by establishing a new objectivity – information – in much the same way that Newton built the physics that bears his name on a new objective instance – force. Just as Newton was unable to define force, so the pioneers of cognitive science were unable to define information, and in both cases the problem of foundational grounding, though transformed, never entirely disappeared. Just as force cannot be reduced to the conceptual resources available in Newton's time, neither can information be reduced to its colloquial meaning, past or present. Information sets the 'level of abstraction', much like force in classical physics or work in thermodynamics" ([9], p 108).

bridge the divide between these traditions, or at least narrow it. One might say that the essence of CR resides in this very insight: a reconceived notion of contextuality.

3. Contextual probability

Contextuality is currently the focus of intensive research in quantum physics and its foundations, where it is closely intertwined with the notion of probability. Given the inherently probabilistic nature of quantum theory,[4] Khrennikov's QLM adopts a contextual probabilistic framework.

Earlier Khrennikov introduced the universal concept of "contextual probability," applicable not only in physics: «*The principle of contextual relativity of probabilities*. Contextuality means that all probabilities depend on complexes of physical conditions, S, – contexts: P(E) ≡ P_s (E). It is meaningless to speak about probability without to determine complex of physical conditions» ([11], pp. 8, 9). Contextual probability refers to the probability—conditional on the available data—of obtaining a particular result in the measurement of a system's quantity. In Khrennikov's framework, such data constitute the *context*, or the experimental conditions [12]. (The standard notion of conditional probability on an event can be viewed as a special case of contextual probability.) On this basis, Khrennikov proposed his *contextual interpretation* of quantum mechanics, known as the "Växjö interpretation." In the simplest case, as is well known, quantum probability is derived from the wave function via Born's rule; in Khrennikov's view, the wave function describes the experimental context itself. Quantum probability, then, is a form of contextual probability.

Independently of Khrennikov's approach, we introduced a similar notion of contextual quantum probability in Ref. [13], as an application of Timothy Williamson's concept of evidential probability [14] to quantum mechanics. Evidential probability is the probability of an event conditional on the entirety of relevant (i.e., contextual) evidence. From the perspective of Williamson's knowledge-first epistemology (KFE), *evidence = knowledge*. Thus, this form of probability is conditional on all relevant contextual knowledge—unlike the subjectivist Bayesian approach. In Khrennikov's sense, context or data correspond precisely to such knowledge or evidence.

However, unlike Khrennikov, we maintain that the wave function encodes knowledge *about* the quantum system (within a context), rather than describing the context itself. As we have argued elsewhere, CR is compatible with KFE and, indeed, serves as its metaphysical foundation [4]. In this respect, our approach to probability is philosophically more fundamental than Khrennikov's, particularly in how it conceptualizes *context* and *data*.

4. QM as a contextual theory: QLM and CQR

QM is both a probabilistic and a contextual theory. However, as with probability, the interpretation of contextuality varies depending on underlying philosophical (pre)assumptions. The most straightforward reading of quantum

[4] However, the evolution of the wave function, as governed by the Schrödinger equation, is deterministic. As for deterministic interpretations of quantum mechanics, they are beyond the scope of our present discussion.

contextuality is operational: The outcome of a measurement depends on the sequence in which physical quantities are measured.[5] In Bell's sense, contextuality—as expressed in quantum correlations—refers to the dependence of the measurement outcome of a variable B on the outcome of another, commuting variable A. Across different contexts (i.e., different values of A), B assumes different values. While quantum correlation is typically understood as a manifestation of non-locality, Khrennikov rejects this interpretation and instead attributes such correlations to contextuality. Independently, we arrived at a similar account within the framework of CR, or more specifically, CQR [2, 15]. This example underscores the fact that in the foundations of physics, philosophical perspectives are inseparable from theoretical—both physical and mathematical—formulations.

The philosophical interpretation of QM profoundly shapes our understanding of quantum phenomena and core concepts such as correlation, nonlocality, superposition, uncertainty, complementarity, entanglement, wave function collapse, probability, and measurement or observation.

In earlier work, we proposed an interpretation of QM grounded in CR, which we termed "Contextual Quantum Realism" (CQR).[6] According to CQR, quantum measurement is not a standalone physical event but rather a contextual phenomenon—a "transition" into a particular epistemic framework. No physical process underlies this shift; rather, we find ourselves situated within one context or another. Outside of a defined context, the notion of measurement becomes meaningless. In this light, the so-called measurement problem—whether in terms of wave function collapse or duality of quantum laws—does not arise.

CQR treats contextuality as a fundamental epistemological and ontological feature of reality. In this light, CQR presents itself as a universal theory of reality—a unifying perspective across the sciences. On this view, scientific theories—particularly QM—function as Wittgensteinian rules or norms ("W-rules") for measuring and engaging with reality within the language games (i.e., phenomena) in which they are applied. CQR thereby accounts for the applicability of the quantum formalism across a diverse range of research domains [2–6] and serves as the philosophical foundation for QLM.

It is important to emphasize that QLM is not merely an application of the mathematical formalism of QM—that is, the theory of operators in Hilbert space—but rather an application of its *partially interpreted* apparatus. Core quantum concepts such as system state (represented by a vector in Hilbert space), observables (as

---

5 Non-commuting observables (e.g., position and momentum) cannot simultaneously possess definite values. As a result, the outcome of their measurement depends on the order in which the measurements are performed. A direct generalization to the field of psychology can be formulated as follows: Responses to questions depend on the order in which they are presented—a phenomenon known as the order effect. Within the framework of QLM, each question is treated as a quantum observable, represented by an operator in a Hilbert space. Asking a question corresponds to measuring this observable, and the response constitutes the measurement outcome. Psychological experiments support the applicability of this quantum-theoretic scheme.

6 As Wittgensteinians in the broad sense, we are cautious about employing the term "interpretation." Our CQR, insofar as it functions as an "interpretation" of QM, should not be viewed as an arbitrary theoretical construct, but rather as a critical and therapeutic attempt to understand QM correctly. The term "interpretation" is therefore used in an extended, non-doctrinal sense. Alternatively, one might speak of CQR as a conception, an understanding, or a point of view. In this light, CQR does not constitute a doctrine, but a philosophical orientation.

Hermitian operators), measurement/observation, and probability are retained. The crucial distinction lies in the nature of the systems under investigation: In the general case, QLM does not concern physical systems, but rather social, human, ecological, economic, political, or other domains. The difference, therefore, is ontological. However, QLM, as an informational approach, abstracts away from ontological specifics. From the standpoint of quantum information theory, there is no essential difference between applying quantum formalism to the microphysical realm and applying it to other fields.[7] In this sense, the world—as a conceptualized and normatively structured dimension of reality—and even reality *as such* may be regarded as quantum. Any ontology, insofar as it is context-sensitive, can be viewed as a *generalized* quantum ontology. The universality of QM thus does not consist in reducing all domains to microphysics, but rather in the fact that microphysics holds no foundational or privileged status; it is merely one domain among others. QM is universal because it is contextual. This explains why, even within physics, both QM and QFT are often regarded not as complete theories, but as open-ended theoretical frameworks. From this perspective, even classical mechanics can—and arguably should—be interpreted through the lens of QM, as possessing a context-sensitive ontology [17, 18].

CR may be understood as a middle path between standard (absolutist) realism and anti-realism. Broadly speaking, realism posits that something exists independently of the subject, whereas anti-realism either denies this or maintains that "reality" is subject-dependent. The former prioritizes externally situated entities, the latter highlights the cognitive modalities by which they are accessed. Both, however, make claims about (pre)determined entities—and for this reason, neither adequately addresses the categorical distinction between the ideal and the real, that is, normativity and contextuality.

CR introduces a crucial distinction: ideal entities—such as norms, concepts, and rules—are objective but not real. Real entities arise through the contextual realization of the ideal ones. The use of a norm or concept is not predetermined; it is embedded within a specific context. Consequently, the object being measured or identified is also not predetermined but comes into determinate being through contextual articulation. From an anti-realist standpoint, this renders the object constructed or unreal, dependent on the subject. That is, anti-realism mistakes non-predeterminacy for non-reality. However, the object—though not yet identified—was what it was prior to its contextual identification; it simply had no identity as an object. By contrast, realism treats it as already constituted, pre-given in the external world, merely waiting to be discovered—thus conflating reality with predeterminacy. This means that for CR, unlike standard realist or anti-realist positions, reality is not reducible to objectivity, nor the real to the true. Both modern and postmodern frameworks tend to reduce reality to objectivity. While some forms of "new realism" emerging in the twenty-first

---

[7] We distinguish between two concepts of information: (1) information as a form of reality with its own specific nature, and (2) information in the epistemic sense—as knowledge.
Information in the epistemic sense is ideal, not part of reality itself. The idealist information doctrine—according to which everything comes from a bit ("it from bit") or the universe is seen as a computational structure—conflates these two notions. This conceptual conflation is also evident in informational interpretations of quantum mechanics. In particular, we have argued that A. Zeilinger's so-called *information principle* is an idealist construct that should be replaced with the more grounded, realist principle of contextuality [16].

century reject this reduction, they often continue to conflate the categories of the ideal and the real—thus remaining, at least partially, within the paradigm of (post) modernity.

So, in our usage, indeterminacy refers simply to the absence of determinacy. It is synonymous with underdetermination or the lack of predetermination. The Heisenberg's uncertainty principle is not merely epistemic, but semantic and, by extension, ontological and epistemic. It is meaningless to assert that a quantum particle possesses both a definite position and momentum—that is, that it occupies a determinate point in classical phase space. This or that indeterminacy entails a corresponding ontological commitment. For example, a particle may have a definite position while its momentum remains completely indeterminate—equally likely to take on any value. Such is the nature of this element of quantum physical reality. It makes no sense to speak of any deeper quantum ontology (or nature). Another manifestation of quantum physical reality is expressed by a particle possessing a well-defined momentum, yet having an entirely indeterminate position. And so on. This constitutes a contextualized quantum ontology—one defined by quantum theory, which is primary in relation to it, yet does not construct it. In this sense, it may be described as an ontology articulated through scientific theory and illuminated by CR, rather than merely in the Quinean semantic sense, where existence is reduced to being the value of a bound variable.

It also follows from the foregoing that Bohr's principle of complementarity is, in essence, a principle of contextuality.[8] Complementary descriptions arise in distinct experimental or conceptual contexts, and complementarity itself may be understood as a manifestation of aspectual change. Thus indeterminacy (non-pre-determinacy or underdeterminacy), contextuality, normativity, complementarity, and the phenomenon of aspectual change—or Gestalt switch[9]—are intimately interconnected. QM and QFT may be regarded as "indetermination theories," normative frameworks (theories as W-rules or norms), or contextual theories. This, we contend, constitutes the very essence of quantum theory—and this is what we deal with in the social sciences and humanities.

Since contextuality—or uncertainty—is the fundamental feature of QM, and is particularly pronounced in the social sciences and humanities, we should expect core quantum phenomena such as superposition, entanglement, correlation ("nonlocality"), "collapse," ontological uncertainty, quantum probability, and others to manifest in these domains as well. In a certain sense, the social sciences and humanities may even appear more "quantum" than quantum mechanics itself. Similar interpretative structures can also be found in classical macroscopic physics, particularly in

---

[8] The principle of complementarity was developed in psychology by W. James. Bohr transferred this principle to CM.

[9] The transition from commutative classical mechanics to non-commutative QM can be interpreted as an instance of aspectual change—or a Gestalt switch—in our conceptual grasp of physical reality. One might also say that the two frameworks complement one another, each emerging from distinct scientific contexts. In this case, the role of context is embodied by the respective "forms of life" of the theories—that is, the established practices and norms governing their application. Ultimately, the shift from classical mechanics to QM can be viewed as a generalization of the theoretical framework understood as a W-rule: a move from one rule to a more encompassing and structurally different one. Grasping this conceptual leap requires engagement with Wittgenstein's problem of rule-following—specifically, how the application and evolution of a rule is conditioned by its contextual enactment ([2], Ch. 6).

classical optics.[10] These are universal phenomena precisely because they emerge from contextuality. In particular, the formalism of QM can be regarded as an interpreted mathematical apparatus for analyzing and navigating situations of uncertainty—particularly in modeling decision-making under such conditions. Empirical evidence increasingly supports this view.

There are, of course, important differences between QM and the QLM applied in the socio-humanities. Outside the domain of quantum physics, Planck's constant plays no role. Accordingly, uncertainty is not formulated in terms of the Heisenberg uncertainty relation—for instance, the principle that product of the uncertainties in position and momentum cannot be smaller than one-half of the reduced Planck constant.

The Copenhagen interpretation of QM is generally classified as an epistemic interpretation. In particular, it is argued that for Niels Bohr, QM is not a theory about what is—that is, about the intrinsic structure of the world—but rather about how we acquire knowledge of the world, about the outcomes of observation. An epistemic interpretation is thus opposed to an ontic one.

A clarification is necessary here. Observation results do not exist in themselves; they inform us about something—namely, they reveal how the world is organized. Knowledge is always *about* something; it is inherently factive (if a subject knows that p, then p). Thus, knowledge cannot be separated from that of which it is knowledge. Consequently, a theory of knowledge—of information or observational results—that is not simultaneously a theory about reality, about the world itself, lacks coherence.

Some informational interpretations of QM are radical in this respect, positing that information is ontologically primary and that QM is fundamentally a quantum theory of information. But this raises a crucial question: *information about what*? If we are dealing with an ontological conception of information, it must be clearly distinguished from the epistemic conception—that is, information understood as arising from observation or measurement.

According to CQR, QM is not just a formalism or an information-processing framework; it is a form of knowledge–knowledge about the world, about quantum reality.

Bohr did not deny the reality of atoms or elementary particles. Rather, his position can be interpreted as giving epistemological primacy over ontology in the ontic sense. Although Bohr does not explicitly use the term "context," his approach is inherently contextual: The role of context is fulfilled by the experimental conditions of a quantum measurement. The preparatory conditions of the experiment constitute its context, and these are uniquely specified by the wave function—a vector in Hilbert

[10] It is classical optics—in which, as it turns out, phenomena such as uncertainty, superposition, entanglement, and nonlocality (quantum-like correlations) also occur, rather than QM—to which Patra turns, proposing a "classical optical modeling of cognition" or a "quantum-like framework" that aspires to provide a philosophical foundation for QLM [19]. One of the foundational elements of his approach is the Hilbert space structure of classical polarization optics. The Indian scholar rightly associates the applicability of QLM with situations of ontological underdetermination, indeterminate choice and insufficient information. As "meta-principles," he adopts the principles of contextuality, complementarity, uncertainty, and nonlocality.
Let us note two fundamental differences between our CQR and Patra's approach. First, Patra's framework is a neo-Kantian one. In particular, he unambiguously interprets Bohr's position as neo-Kantian. Second, Patra treats quantum probability as subjective probability. Finally, from the perspective of CQR, the proposed meta-principles are instantiations of a single underlying principle of contextuality.

DOI: http://dx.doi.org/10.5772/intechopen.1012092

space. (Accordingly, as Khrennikov emphasizes, the wave function describes the experimental context; what is typically referred to as the state of a quantum system is, in fact, its context.)

The conditions in which a system finds itself after an experiment—resulting in the "collapse" of the wave function and the observation of a specific value of a measurable physical quantity—constitute a new context. According to CQR, the result of measurement is not predetermined but becomes identifiable within this emergent context, which is itself the context *of* the result. Ontology, then, is context-sensitive: A given outcome does not exist independently, but only within a specific contextual configuration.

From the perspective of CQR, QM is an epistemic theory in the sense that it functions as a W-rule or normative structure—prior to ontology. Even if QM is regarded merely as a formal framework, it nonetheless concerns reality. (Analogously, several contemporary analytic philosophers of science distinguish between "theory-first," "epistemology-first," and "mathematics-first" approaches—such as that of D. Wallace—and the contrasting "ontology-first" orientation.)

As previously discussed, CQR—or more broadly CR—seeks to reconcile Bohr's epistemological (often labeled "anti-realist") standpoint with Einstein's realist commitments ([5], p. 222). QM, construed as a W-rule, also aligns with Einstein's notion of a principle theory. (For Einstein, it is the theory that tells us what is observable).

Khrennikov frames QLM at the epistemic level. At the same time, he remains a realist, affirming the existence of an ontological (ontic) domain—though QLM itself does not attempt to model this level. QLM concerns the processing of information, and information properly belongs to the epistemic domain.

From the standpoint of CR, one could also argue that QLM implies the existence of a deeper—conceptually, though not metaphysically—ontological (not ontic) level of reality: a pre-objectivized stratum. This is, in essence, the domain of fundamental indeterminacy.

From the perspective of QLM, the existence or non-existence of "hidden variables" is largely irrelevant—since, in either case, they remain hidden by virtue of quantum mechanics being an epistemic theory. The Copenhagen interpretation denies the existence of hidden variables, whereas Bohmian mechanics assumes their existence (nonlocal hidden variables can, at least formally, be introduced into QM).

In the socio-humanitarian context, only certain variables are relevant to the model; others may be considered "hidden," yet they remain accessible—at least in principle—if one chooses to observe them. Khrennikov points to this distinction as a key difference between QM and the socio-humanities. Yet we observe a deeper affinity between the two: in both domains, ontology is context-sensitive. What is ascribed to hidden variables does not define ontology in the ontic sense—neither in QM nor in socio-humanitarian studies. The symmetry between them suggests that, in quantum mechanics as well, there may exist a conceptually deeper ontological (as opposed to ontic) level of reality itself.

Khrennikov has criticized mathematicians for their lack of interest in the nature of probabilistic events, treating them instead as objectively random. Yet to engage with the ontological (not ontic) level is precisely to open the question of what randomness means. A random event, understood contextually, is neither predictable nor predetermined—yet this does not imply that, at the ontological (not ontic) level, it lacks a cause.

A rough analogy can be drawn with Kant's distinction between the world of phenomena and the world of things-in-themselves. However, from the standpoint of CQR, the world of genuine phenomena coincides with the world of things in themselves, and the concept of a thing in itself is meaningless unless it is treated as the concept of a non-conceptualized thing.

Post facto, a random quantum event—such as a particular measurement outcome—can be given a causal explanation. A similar position appears to have been adopted by the German philosopher and physicist Hermann Grete, who aimed to preserve causality in QM within a neo-Kantian framework [20]. For her, the unpredictability of individual quantum events is compatible with the existence of causes and their retrospective, classically causal explanations. CQR is broadly aligned with this view, though it diverges from the neo-Kantian philosophical foundation.

5. QLM-formalism

QLM can be formulated using several mathematically equivalent frameworks: quantum probability theory (or its generalized forms), the formalism of non-commutative operators in Hilbert space,[11] quantum logic (i.e., non-Boolean (non-distributive) algebra), and quantum information theory.

In standard QM, the postulates of the theory are typically stated in the language of operator theory on Hilbert space, with quantum probability introduced only afterward. Khrennikov, however, proposed a reversed approach: He derives both the Hilbert space formalism and Born's rule from quantum probability theory itself—understood as a generalization of classical probability that includes an interference term (i.e., the additivity of probabilities is violated).[12] Quantum probability, for example, accounts for the interference pattern observed when quantum particles pass through two slits.

The relationship between QM and classical mechanics can be viewed as a generalization: QM arises from replacing commutative observables (real-valued quantities) with non-commutative ones (operators), or, put differently, from replacing a phase space with commutative coordinates by a non-commutative space. The quantum commutator is the analogue of the classical Poisson bracket. In fact, all of non-relativistic QM can be derived from a single fundamental expression: the commutator of position and momentum (analogous to their classical Poisson bracket) is proportional to Planck's constant. In particular, the Heisenberg uncertainty relation follows from the fact that the simultaneous exact measurement of position and momentum is fundamentally meaningless.

The non-distributivity (though not non-commutativity) of quantum logic—unlike classical logic, which is distributive—corresponds to the non-commutative structure of operators in Hilbert space. Within the Hilbert space formalism, there is

[11] While quantum physics uses both finite-dimensional and infinite-dimensional Hilbert spaces, Khrennikov uses only finite-dimensional spaces, that is, linear algebra, to model macroscopic phenomena outside quantum physics. In the simplest case, this is a two-dimensional space.

[12] More precisely, for Khrennikov, the difference between classical and quantum probabilities is in the so-called "total probability formula." On the contrary, R. Feynman believed that quantum probability differs from classical probability by the fact that it violates the additivity condition.

not a single Boolean algebra but rather a family of incompatible yet complementary Boolean algebras that together form a non-Boolean structure.

In quantum information theory, the notion of the qubit (quantum bit) is introduced. In classical information theory, the minimal unit of information is a bit: We have minimal information when we know which of two possibilities obtains – for instance: "yes" or "no," "0," or "1," "the spin projection on the z-axis is +½" or "–½," the proposition *p* is true or false, etc.

A qubit is a superposition of the states "0" and "1." Is it therefore a new unit of information—the smallest *quantum* unit of information? Does quantum mechanics transform our concept of information? We do not think so. One should distinguish between information in the epistemic sense and "information" in the material/physical (ontological) sense as a carrier of information in the epistemic sense. For example, the projection of spin onto the z-axis (±½) is "information" in the physical sense—it functions as a *carrier* of epistemic information, which can be formulated propositionally.

A classical bit can be understood both epistemically and physically. By contrast, a qubit, as the quantum state of a minimal (elementary) system, is a unit of physical information. Only through measurement does it yield epistemic information, which is expressed in classical bits.

6. Beyond physics

Thus, the structure of QM can be represented by the following triple: an algebra of observables (O), a Hilbert space (H) on which the operators of this algebra act, and a state (ϕ)—a vector in the space H: (O, H, ϕ). This structure can be extended to the domain of social sciences and humanities with the corresponding elements indexed as "s" for "social and humanities": (O_S, H_S, ϕ_S). Only empirical investigation can reveal when this structure is applicable, when it is not, and when it requires modification.

As previously mentioned, phenomena such as the order effect and the psychological principle of complementarity illustrate these extensions. Khrennikov speaks of quantum-like effects, referring to phenomena and systems in which quantum features manifest and are empirically observable. He defines quantum-like systems as information processors governed by the laws of quantum probability and information [21]. However, from the perspective of CR, one may speak of genuine quantum effects and phenomena beyond the ontology of quantum microphysics—genuine in the sense that they are not derivative or mimetic. By the same token, one could characterize quantum phenomena as "social-like" within the quantum domain. The essential difference, as already noted, lies in the presence of a fundamental physical constant in quantum physics—the Planck constant.

Earlier, drawing on the critical work of M. Gabriel and J. Benoist, we concluded that what distinguishes human beings from artificial intelligence is the presence of the ideal *Geist*, understood as normativity [22]. Andler articulates a similar distinction: "Intelligence is not a thing, a phenomenon, a process or a function, but a norm that applies to behavior: it qualifies the relation between human and her world, and in a way that is never objective and final" [23]. Human intelligence, then, is primarily the sensitivity to context—a normative orientation toward a particular situation and the capacity to make optimal decisions under uncertainty. In contrast, artificial intelligence excels at solving well-defined problems but lacks autonomy. Given that

QM is a fundamentally contextual theory, one might suppose that a human-like form of artificial intelligence would be grounded in quantum logic rather than classical logic [24]. It must be emphasized, however, that such AI would not embody genuine intelligence but rather a simulation thereof.

This reinforces our thesis that CR (CQR) offers a robust philosophical foundation for QLM, particularly in modeling human behavior and consciousness.[13]

Quantum concepts have also increasingly found their way into socio-political theory, especially in areas such as international relations, global politics, and security studies. Yet, as we have argued, what many scholars describe as the "quantum paradigm" in these disciplines often constitutes less a genuine departure from the outdated Modern framework and more a postmodernist radicalization of it. In contrast, the conceptual scheme of CQR offers a substantive and authentic quantum alternative [26].

In the following paragraph, we illustrate the foregoing by examining the position of A. Wendt.

7. A. Wendt's position

A. Wendt proposed incorporating subjectivity and consciousness into the social sciences, and advocated for a unification of physical and social ontology within an anthropological and quantum perspective. He writes: "Subjectivity – and here I mean conscious subjectivity – can and should be 'brought back in' to social science" ([27]. p. 285). The author also claims that social reality is quantum-mechanical ([27], p. 36). He grounds his approach in an interpretation of quantum mechanics informed by neutral monism and accordingly proposes a naturalistic panpsychic "quantum theory of consciousness," according to which humans are "walking wave functions" (i.e., superpositions of possibilities/potentialities), and social structures are superpositions of mental states ([27], p. 283). For Wendt, "the quantum argument is in fact very simple: consciousness is an aspect of matter at the elementary level (panpsychism), which is amplified upward by quantum coherence in the brain" ([27], p. 292). The collective arises through the entanglement of individual consciousnesses.

Independently, we—like Wendt—also proposed a unified, contextual perspective on both the natural and social sciences, and argued that social reality is in fact "quantum" reality [17, 18]. This refers to the CQR mentioned above. For CQR, reduction and correlation are understood as transitions into a corresponding context—be it a reduced state or a specific correlation—accompanied by a transformation in the state of information and, accordingly, the identity of the object under consideration, rather than as physical processes. In other words, quantum mechanics points to the context-sensitivity of ontology (of real objects), whereas Wendt maintains that it compels us to adopt a "flat ontology," in which all entities exist on equal ontological footing.

Since the conceptual framework and the physico-mathematical structure of quantum mechanics account for contextuality, quantum epistemology proves to be universal. The general structure of quantum mechanics functions as a W-rule—a means of measuring reality within contextual language games (i.e., normative practices) of its application [17, 18]. CQR does not posit a fixed metaphysical ontology.

---

[13] On the modeling of consciousness and the "hard problem" of consciousness in the context of QLM, see, for example [25].

Instead, it treats primary sensory experience—prior to any division into the physical and the mental—as reality. Physical and mental realities are regarded as secondary, in the sense of their determinateness (givenness). In neutral monism, a fundamental neutral ontology is introduced instead.

Many of Wendt's claims bear superficial resemblance to ours, yet his philosophical premises are markedly different. For example, Wendt maintains that language is an instrument of expressive measurement; measurement is a speech "act," an act of using language within a context, through which potential meanings become actualized (a collapse of a superposition of meanings occurs), and quantum interference and entanglement are manifested ([27], p. 217).

That being said, Wendt contends that measurement—or the use of language—affects that which is being measured, and that the object of measurement did not exist prior to measurement. He writes: "In most interpretations of quantum theory particles cannot be said to exist prior to measurement, and in preparing quantum systems for observation an entanglement is created with the observer that affects what is eventually seen. That does not mean that the observer literally creates reality, but it does mean that she participates in what is actually observed, and as such observation cannot even in principle approximate the classical ideal of separation" ([27], p. 36). "In quantum physics objects cannot be said to exist prior to their realization in measurement by an observer. (...) The collapse of the wave function is a process (...) in which not only a material particle is produced, but also a sub-atomic experience of that process" ([27], p. 285).

These formulations are either false or ambiguous and open to varied interpretation. From the perspective of CQR, however, they can be read as tautologies: By definition, objects as objects—in particular, particles as particles (rather than waves)—presuppose their identification *via* measurement or observation. In the process of measurement/observation, objects (such as particles or waves) arise as objects, but not as physical things. According to CQR, it is incorrect to assert that what is measured does not exist prior to measurement. The use of language—as a tool of observation and a W-rule (norm)—does not affect the thing being measured, but instead identifies what is and has already been prior to measurement.

At the same time, CQR holds that the measurable entity, prior to its measurement, lacked identity—it was not (pre)determined and did not yet constitute an object. In this sense, the claim that objects did not exist prior to their realization or observation can be interpreted as a grammatical (logical) truth, rather than a substantive one.

According to Wendt, quantum mechanics—and by extension, quantum social science—undermine the subject-object dichotomy and subject-object dualism in the following sense:

"In asking a given question: (1) social scientists are collapsing a socially shared wave function, locally for us as researchers, and thereby calling forth a reality rather than just passively observing one given to us as spectators, and (2) our research is therefore responsible, at the micro-level, for helping to create, sustain and/or transform that reality. This is not to say that if social scientists refuse to study something which they disapprove of – as some in IR [international relations] would have us do with the state – it will disappear, since most social facts are sustained by thousands or millions of lay people 'observing' them, not by social scientists. But it does emphasize our complicity in that process, both empirically and politically.

These quantum effects suggest that the hope of social scientific realists – that through the continuing improvement of our theories and measurement techniques we will get closer and closer to truths of social life – is misplaced" ([27], p. 287).

This, once again, is an idealist position that conflates the conceptualization (identification) of a particular slice of reality with the creation of reality itself, neglecting the logic of phenomena. According to CQR, every phenomenon—quantum or classical—by definition presupposes an observer (something appears to someone). Like Wendt, CQR rejects the subject-object dichotomy, but not the distinction itself, which is logical in nature. The (real) object belongs to the category of the real. The subject is dual: On the one hand, it is part of reality (as a concrete, real subject), and on the other, when functioning as an observer, it performs a purely logical role, shifting its status to the ideal.

According to CQR, we come to know the things themselves; Whereas for Wendt, as for Kant, the things themselves are unknowable—there are inherent limits to human cognition: "The experience of a particle is ephemeral, and 'what it is like' to be one is probably beyond the human capacity to understand. Indeed, we face such an epistemic limit even in our dealings with other organisms, like Nagel's bat, whose experiences are not ephemeral but persist in memory. At the highest level of biological organization – apes, chimps, and dogs – we might gain some genuine second-person understanding, but in general when it comes to interpreting meanings we are stuck in a bubble with our fellow human beings. Note, however, that these are limits to our knowledge, and as such do not mean that other organisms do not have meaningful experiences, much less that they are mere machines or objects" ([27], p. 285). Wendt's position, therefore, remains within the boundaries of the modern paradigm.

8. Conclusion

We have argued that the formalism of QM can be applied to modeling phenomena and systems beyond QM itself and even beyond physics—particularly in the domain of social sciences and humanities. Such applications are well documented (e.g., Khrennikov's QLM). Additionally, we have substantiated the thesis that CR/CQR can provide a philosophical foundation for QLM. CR/CQR integrates conceptual therapy with the metaphysics of reality, the ideal, and of things themselves. It is a meta-meta-theory that enables the unification of phenomena of vastly diverse nature—from the exact and natural sciences to the humanities.

Conflict of interest

The authors declare no conflict of interest.

Author details

Francois-Igor Pris
Institute of Philosophy of the National Academy of Sciences of Belarus, Minsk, Republic of Belarus

*Address all correspondence to: frigpr@gmail.com

References

[1] Khrennikov A. Open Quantum Systems in Biology, Cognitive and Social Sciences. Cham: Springer; 2023. 371 p

[2] Pris IE. Contextuality of Ontology and Contemporary Physics. St. Petersburg: Aletheia Publications; 2020. 354 p

[3] Pris IE. Higgs Boson, Quantum Strings, and Philosophy of Physics. St. Petersburg: Aletheia Publications; 2021. 192 p

[4] Pris IE. Knowledge in Context. St. Petersburg: Aletheia Publications; 2022. 720 p

[5] Pris IE. Contextual Quantum Realism and Other Interpretations of Quantum Mechanics. Moscow: URSS Publications; 2023. 304 p

[6] Pris IE. Morality, Quantum Mechanics, Religion, Technology, Artificial Intelligence through the Lens of Contextual Realism. Moscow: Lenand Publications; 2024. 280 p

[7] Patra S, Yeddanapudi S. Quantum-like contextual utility framework application in economic theory and wider implications. In: Chakraborti A, Haven E, Patra S, Singh N, editors. Quantum Decision Theory and Complexity Modelling in Economics and Public Policy. Cham: Springer; 2023. 259 p

[8] Benoist J. Toward a Contextual Realism. Cambridge, Massachusetts, London: Harvard UP; 2021. 216 p

[9] Andler D. Le silhouette de l'humaine. Paris: Gallimard; 2016. 555 p

[10] Benoist J. L'adresse du réel. Paris: Vrin; 2017. 376 p

[11] Khrennikov A. Växjö interpretation of quantum mechanics. In: Proc. Int. Conf. "Quantum Theory: Reconsideration of Foundations". Ser. Math. Modelling in Physics and Cognitive Sciences. Vol. 2. Vaxjo: Vaxjo University Press; 2002. pp. 163-170. Available from: https://arxiv.org/abs/quant-ph/0202107 [Accessed: June 10, 2024]

[12] Khrennikov A. Contextual Approach to Quantum Formalism. Springer; 2009. 384 p

[13] Pris IE. On quantum probability. Philosophy of Science (Novosibirsk). 2022;**3**(94):46-65. DOI: 10.15372/PS20220305

[14] Williamson T. Knowledge and its Limits. Oxford: Oxford UP; 2022. 352 p

[15] Pris IE. On quantum causality. Vestnik of Saint Petersburg University. 2023;**3**:462-477. DOI: DOI.org/10.21638/spbu17.2023.305

[16] Pris IE. On a fundamental conceptual principle of quantum mechanics. Philosophy of Science (Novosibirsk). 2021;**4**(91):82-92

[17] Pris IE. Humanities and quantum mechanics. Philosophy of Science (Novosibirsk). 2020;**2**(85):113-130. DOI: 10.15372/PS20200208

[18] Pris IE. Social reality – "Quantum" reality. Dialogue. 2020;**2**:8-11

[19] Patra S. Why Quantum-like Framework for Economic Theory and Social Sciences? 2023. Available from: https://www.researchgate.net/publication/371170354_Why_quantum-

like_framework_for_Economic_theory_And_social_sciences [Accessed: June 04, 2024]

[20] Soler L. The convergence of transcendental philosophy and quantum physics: Grete Henry-Hermann's 1935 pioneering proposal. In: Bacciagaluppi G, Crull E, editors. Grete Hermann – Between Physics and Philosophy. Dordrecht: Springer; 2017. pp. 55-69

[21] Khrennikov A. Quantum-like models in biology, cognition, and AI. In: The 24th Växjö Conference on Quantum Foundations Quantum Information and Probability: From Foundations to Engineering (QIP24), June 11-14, 2024. Växjö, Sweden: Linnaeus University; Forthcoming

[22] Pris IEM. Gabriel's neo-existentialism against artificial intelligence. In: Kruglikov SV, Kasavin SN, editors. II Forum of IT-Academgorod «Artificial Intelligence in Belarus». 12-13 October 2023. Minsk, OIIP: NAS of Belarus Publications; 2023. pp. 208-216

[23] Andler A. Intelligence artificielle, intelligence humaine: la double énigme. Paris: Gallimard; 2023. 432 p

[24] Khrennikova p Human centered AI for financial decisions. In: The 24th Växjö Conference on Quantum Foundations Quantum Information and Probability: From Foundations to Engineering (QIP24), June 11-14, 2024. Växjö, Sweden: Linnaeus University; Forthcoming

[25] Baladrón C, Khrennikov A. Outline of a microscopic physical model of elementary consciousness. In: The 24th Växjö Conference on Quantum Foundations Quantum Information and Probability: From Foundations to Engineering (QIP24), June 11-14, 2024. Växjö, Sweden: Linnaeus University; Forthcoming

[26] Pris IE. On quantum ontology in social and political sciences. Voprosy filosofii. 2025;**3**:5-13

[27] Wendt A. Quantum Mind and Social Science: Unifying Physical and Social Ontology. Cambridge: Cambridge University Press; 2015. 354 p

Section 2

Quantum Dynamical Simulation

Chapter 3

Performance Analysis of Nonequilibrium Quantum Otto Engines in Relation to Quantum Coherence

Jeolous Malamula Nyasulu, Chunyang Wang, Zengxuan Zhao, Mushtaq Rana Imran, Elson Rodger Katete, Akuonani Zakeyo Phiri and Evans Kumalo Mtambo

Abstract

In this chapter, we examine how quantum coherence affects the effectiveness and performance of quantum Otto engines running in nonequilibrium conditions. Energy conversion on a quantum scale can be understood through the use of quantum Otto engines, which employ cyclic thermodynamic processes involving isochoric and adiabatic transformations. We examine how the ability to extract work from thermal reservoirs at different temperatures is impacted by preserving quantum coherence among the engine's operating states. We investigate the dynamics of coherence under various nonequilibrium conditions and configurations, including changing temperatures and coupling strengths, using the Quantum Master Equation framework. According to our research, engines with improved quantum coherence perform noticeably better than their classical counterparts, allowing for higher efficiencies and more work extraction. We also explore optimization techniques and real-world applications for creating quantum heat engines that make use of coherence as a vital resource. The potential of coherent quantum states in powering next-generation energy systems is highlighted in this work, which connects theoretical developments in quantum thermodynamics with real-world applications.

Keywords: quantum Otto engine, quantum coherence, nonequilibrium thermodynamics, engine performance, energy conversion, quantum heat engines

1. Introduction

Over the past 10 years, there has been a notable surge in the study of quantum thermodynamics, indicating a promising convergence of thermodynamic and quantum mechanical concepts. The quantum Otto engine stands out among the other models studied because it has the potential to outperform classical engines by taking advantage of the quirks of quantum coherence. In order to identify the fundamental mechanisms that improve engine efficiency under nonadiabatic processes, this chapter explores the complex relationship between quantum coherence and the performance of nonequilibrium quantum Otto engines.

A characteristic of quantum systems, quantum coherence permits particles to exist in superpositions of states, potentially improving thermodynamic cycle performance. Prior research has mostly concentrated on the function of quantum coherence in equilibrium situations, demonstrating how coherence can increase productivity and efficiency [1, 2]. However, there is still a dearth of research on the performance analysis of coherence in nonequilibrium settings of quantum Otto engines, which this chapter aims to fill.

When the body of existing literature is critically examined, a pattern of research emerges that frequently restricts the analysis to idealized systems or equilibrium conditions. The foundation for quantum heat engines, for example, was established by [3, 4]; however, their work did not take nonequilibrium dynamics into consideration. The significance of nonequilibrium processes is emphasized in more recent research, such as that by Plastina et al. [5], but their models do not adequately incorporate the effects of coherence.

Furthermore, while studies like Refs. [6, 7] have successfully demonstrated that dissipative environments can have a significant impact on quantum systems, they are unable to directly connect these environments to the role of coherence in improving engine performance. This discrepancy is important because, in real-world applications, quantum engines frequently experience large nonequilibrium effects under operating conditions, particularly in modern quantum technologies like quantum computing and quantum information processing.

The goal of this paper is to close this crucial gap by carefully examining the use of quantum coherence in nonequilibrium quantum Otto engines [8]. Our objective is to gain a more thorough understanding of the thermodynamic potentials of these engines by extending the current models to include quantum coherence under nonequilibrium circumstances. Our investigation will shed light on both the theoretical and practical ramifications that may influence future quantum thermal machine designs.

2. Fundamental principles of the quantum Otto engine in the thermodynamic cycle

The quantum Otto engine functions as a quantum counterpart to classical heat engines, adhering to thermodynamic principles through a cyclic process involving work extraction and heat absorption [9]. This engine operates *via* four key processes: two adiabatic (where no heat is exchanged) [10] and two isochoric (where volume remains constant) [11]. The analysis of these engines focuses on their efficiency in quantum environments, highlighting how quantum states are manipulated to harness energy in a manner analogous to classical engines.

In the first stage, referred to as the quantum isochoric process, the system absorbs heat Q_H from a heated reservoir at a temperature T_H while maintaining constant volume. At thermal equilibrium, the working material's density matrix is expressed as

$$\rho_h = \frac{\exp\left(-\frac{H'_h}{k_B T_h}\right)}{Z_h} = \sum_n p_n^h |\Psi_n\rangle\langle\Psi_n|, \tag{1}$$

where $p_n^h = \frac{\exp\left(-\frac{E_n^h}{k_B T_h}\right)}{Z_h}$ indicates the population probabilities and $Z_h = \sum_n \exp\left(-\frac{E_n^h}{k_B T_h}\right)$ is the partition function. The second stage involves an adiabatic expansion where the system, isolated from the heat source [12], evolves from an initial to a final state without heat exchange by absorbing energy and experiences a transition in Hamiltonian [13–15] from H'_h to H'_c.

The third stage encompasses a quantum isochoric cooling process where heat Q_C is emitted to a cold reservoir at a temperature T_C while maintaining constant volume. The density matrix is updated during this stage to

$$\rho_c = \frac{\exp\left(-\frac{H'_c}{k_B T_c}\right)}{Z_c} = \sum_n p_n^c |\Psi_n\rangle\langle\Psi_n|, \tag{2}$$

with $Z_c = \sum_n \exp\left(-\frac{E_n^c}{k_B T_c}\right)$. The cycle is completed with adiabatic compression, yielding work from the transition of the system back to its initial state. The quantification of heat transfer is expressed as

$$Q_h = \mathrm{Tr}\left(H'_h \rho_h\right) - \mathrm{Tr}\left(H'_h \rho_c\right) \quad \text{and} \quad Q_c = \mathrm{Tr}\left(H'_c \rho_c\right) - \mathrm{Tr}(H_c \rho_h), \tag{3}$$

with energy conservation, allowing us to calculate the net work output as

$$W = Q_h + Q_c, \tag{4}$$

and the engine's efficiency is defined as $\eta = \frac{W}{Q_h}$.

3. The role of quantum coherence

The phenomenon where quantum states exist in superpositions and may exhibit interference effects is known as quantum coherence [16]. Coherence has been shown to be important in manipulating quantum states for a variety of applications, such as quantum thermodynamics and quantum computing [17]. Koyanagi and Tanimura [18] mainly concentrate on creating a framework for nonequilibrium thermodynamics by employing a system-bath model and the dimensionless minimum work principle. Although "coherence" is not used specifically, it has a subtle influence, particularly in the focus on quantum entanglement and non-Markovian dynamics. Understanding and enhancing the performance of quantum Otto engines is made possible by this implicit presence of coherence. The chapter's emphasis on non-Markovian and nonperturbative techniques subtly recognizes the value of coherence approaches [19, 20]. The memory effects and correlations resulting from the system-bath interaction are inherently neglected by Markovian approximations, which are frequently used to simplify the description of open quantum systems. In essence, these memory effects are an expression of the system's and its surroundings' sustained quantum coherence. The author does not, however, specifically measure or examine the precise role that coherence plays in the engine's performance [21]. Without breaking down the underlying quantum dynamics to separate the effects of coherent superpositions and quantum interference, the emphasis is still on analyzing thermodynamic potentials and work diagrams.

4. Quantum Master Equation framework

We generally use the Quantum Master Equation (QME) to study the dynamics of quantum Otto engines in nonequilibrium conditions. A density matrix's evolution under the influence of an external environment is mathematically described by the QME [22]. The dynamics of a quantum system interacting with an external environment or bath are described by this framework. The Born-Markov approximation and the density matrix formalism are commonly used in this derivation.

Consider a quantum system S described by the density matrix ρ_S and an environment (or bath) E described by the density matrix ρ_E. The total system, which combines both the system and the environment, is denoted by ρ_{SE}.

The total Hamiltonian for the composite system can be expressed as

$$H = H_S + H_E + H_{SE}, \tag{5}$$

where H_S represents the system's Hamiltonian, H_E represents the environment's Hamiltonian, and H_{SE} represents the system-environment interaction Hamiltonian.

For the initial state, we assume that the system and environment's overall state can be factored as follows $t = 0$:

$$\rho_{SE}(0) = \rho_S(0) \otimes \rho_E, \tag{6}$$

where $\rho_S(0)$ is the initial state of the system and ρ_E is the state of the environment.

Time-based development according to Schrödinger's equation, the evolution of the total density matrix under the influence of the total Hamiltonian is as follows [23]:

$$\rho_{SE}(t) = U(t,0)\rho_{SE}(0)U^\dagger(t,0), \tag{7}$$

where $U(t,0) = e^{-iHt}$ is the time evolution operator.

We identify the degrees of freedom of the environment for the reduced dynamics of the system, which concentrate on the dynamics of the system S:

$$\rho_S(t) = \mathrm{Tr}_E[\rho_{SE}(t)]. \tag{8}$$

Substituting the expression for $\rho_{SE}(t)$:

$$\rho_S(t) = \mathrm{Tr}_E\left[U(t,0)\rho_S(0) \otimes \rho_E U^\dagger(t,0)\right]. \tag{9}$$

In the weak coupling regime, the Born Approximation was used. We assume that the interaction Hamiltonian H_{SE} is weak enough that there is no significant change in the environment while interacting with the system [24]. Consequently, we can estimate $U(t,0) \approx 1$:

$$\rho_S(t) \approx \rho_S(0) + \mathrm{Tr}_E\left[-i\int_0^t dt'[H_{SE}, \rho_{SE}(t')]\right]. \tag{10}$$

The dynamics of the system are assumed to be local in time in the Markov approximation, which implies that environmental effects occur instantly. This enables us to estimate the integral over time t' as relying solely on the system's present state. The density matrix evolution takes on the following form when this simplification is applied:

$$\frac{d\rho_S(t)}{dt} = \mathcal{L}[\rho_S(t)], \tag{11}$$

where $\mathcal{L}$ is the Liouvillian superoperator, which encodes the effects of the environment on the system [25].

The Lindblad form can be used to express the Liouvillian using the Lindblad formulation of the master equation:

$$\mathcal{L}[\rho_S] = -i[H_S, \rho_S] + \sum_j \left(L_j \rho_S L_j^\dagger - \frac{1}{2}\left\{ L_j^\dagger L_j, \rho_S \right\} \right), \tag{12}$$

where the term enclosed in curly brackets indicates the anti-commutator, and the Lindblad operators L_j represent the system's interaction with the environment.

The final form of the Quantum Master Equation is:

$$\frac{d\rho_S(t)}{dt} = \mathcal{L}[\rho_S(t)], \tag{13}$$

where $\mathcal{L}$ captures the dissipative effects brought about by the interaction with the environment as well as the coherent evolution caused by the system's Hamiltonian [26]. Analysis of nonequilibrium dynamics and quantum thermodynamic processes is made possible by this equation, which characterizes the time evolution of the density matrix of the quantum system under the influence of its surroundings.

4.1 Derivation of the Quantum Master Equation (QME)

Tracing out the bath degrees of freedom yielded the reduced density matrix of the system $\rho S(t)$, whose time evolution is described by the QME. The following procedures are usually involved in the derivation:

Applying the von Neumann formula: The von Neumann equation describes how the entire system (system plus bath) changes over time:

$$\frac{d\rho(t)}{dt} = -i[H, \rho(t)] \tag{14}$$

Where:

$\rho(t)$ is the density matrix of the total system. $H = H_S + H_E + H_{SE}$ is the total Hamiltonian.

In relation to the free Hamiltonians $(H_S + H_E)$, this changes the interaction picture:

$$\rho_I(t) = e^{i(H_S+H_E)t}\rho(t)e^{-i(H_S+H_E)t} \tag{15}$$

$$H_{SE,I}(t) = e^{i(H_S+H_E)t}H_{SE}e^{-i(H_S+H_E)t} \tag{16}$$

The von Neumann equation in the interaction picture becomes

$$\frac{d\rho_I(t)}{dt} = -i[H_{SE,I}(t), \rho_I(t)] \tag{17}$$

Formal Integration: Integrate the equation from 0 to t:

$$\rho_I(t) = \rho_I(0) - i\int_0^t dt' [H_{SE,I}(t'), \rho_I(t')] \tag{18}$$

Approximation of weak coupling (Born Approximation): Assume that the bath stays relatively close to its initial state because the system-bath interaction is weak. As a result, we can roughly calculate the total density matrix as a product:

$$\rho_I(t) \approx \rho_{S,I}(t) \otimes \rho_E \tag{19}$$

Assuming the bath is in thermal equilibrium, ρ_E represents its density matrix.

We make one iteration by replacing the integrated equation with the product state approximation using the second-order perturbation theory:

$$\frac{d\rho_I(t)}{dt} = -i\left[H_{SE,I}(t), \rho_{S,I}(t) \otimes \rho_E\right] - \int_0^t dt' \left[H_{SE,I}(t), \left[H_{SE,I}(t'), \rho_{S,I}(t') \otimes \rho_E\right]\right] \tag{20}$$

Take a partial trace of the bath's degrees of freedom in order to trace over it:

$$\begin{aligned} \frac{d\rho_{S,I}(t)}{dt} &= \mathrm{Tr}_E\left\{-i\left[H_{SE,I}(t), \rho_{S,I}(t) \otimes \rho_E\right]\right\} \ldots \\ &- \mathrm{Tr}_E\left\{\int_0^t dt' \left[H_{SE,I}(t), \left[H_{SE,I}(t'), \rho_{S,I}(t') \otimes \rho_E\right]\right]\right\} \end{aligned} \tag{21}$$

According to the Markov approximation, the bath has a short memory, which means that the system's evolution at time t is solely dependent on its current state and not on its historical background. This enables us to extend the integration limit to infinity and replace the integral $\rho_{S,I}(t)$ within $\rho_{S,I}(t)$:

$$\int_0^t dt' \ldots \approx \int_0^\infty dt' \ldots \tag{22}$$

When the bath correlation time is significantly less than the system's relaxation time, this approximation holds true.

In order to write the QME in the Lindblad form, Lindblad follows these procedures and applies a few algebraic manipulations:

$$\frac{d\rho_S(t)}{dt} = -i[H_S, \rho_S(t)] + \sum_j \left(L_j \rho_S(t) L_j^\dagger - \frac{1}{2}\left\{L_j^\dagger L_j, \rho_S(t)\right\}\right) \tag{23}$$

where the Lindblad operators, which characterize the decoherence and dissipation processes, are L_j. They rely on the properties of the bath and the interaction between the system and the bath. The anti-commutator is $A, B = AB + BA$.

4.2 System-bath interaction Hamiltonian (HSE)

The interaction between the system and the bath. The interaction between the quantum Otto engine (the system) and its surroundings (the bath) is described by the Hamiltonian (HSE). Since it determines the character of energy exchange and decoherence processes, the particular form of HSE is essential.

$$H_{SE} = \sum_j \sigma_j \otimes B_j \tag{24}$$

Where:

An index called j adds up various interaction terms. σ_j are operators operating on the Hilbert space of the system. Depending on the particular coupling mechanism, these are usually Pauli operators $(\sigma_x, \sigma_y, \sigma_z)$ or combinations of them. They stand for the system's physical characteristics that interact with the bath. B_j are operators operating on the Hilbert space of the bath. Usually, these are stated in terms of the bath modes' creation and annihilation operators:

$$B_j = \sum_k g_{k,j}\left(b_k + b_k^\dagger\right) \tag{25}$$

Where

The index k is a sum of the various modes of the bath, such as the various harmonic oscillator frequencies. The coupling constants $g_{k,j}$ decides how strongly the system operator σ_j and the bath mode k interact. The spectral characteristics of the bath are frequently determined by these coupling constants, which are very important. For the k-th mode of the bath, the annihilation operator is b_k, and the creation operator is $b_k^\dagger$. According to the particular coupling mechanism, they satisfy the commutation relation $[b_k, b_k^\dagger] = \delta_{k,k'}$. They stand for the system's physical attributes that interact with the bath. The operators B_j act on the Hilbert space of the bath. Operators of the bath modes are used to express $(\sigma_x, \sigma_y, \sigma_z)$, which represent various kinds of interactions; for example, $\sigma_z \otimes B_j$ represents an energy exchange interaction. The energy level of the system affects the bath and vice versa. Additionally, a dephasing interaction is presented by $\sigma_x \otimes B_j$ or $\sigma_y \otimes B_j$. While (B_j) is a bath that is frequently represented as a set of harmonic oscillators, the bath affects the coherence of the system, but there is no direct energy exchange. The excitation and de-excitation of these oscillators, which indicate the energy absorbed or released by the bath, are described by the creation and annihilation operators [27]. The strength of the system's interaction with the various bath modes is then determined by the coupling constants $\left(g_{k,j}\right)$. The spectral density of the bath is defined as

$$J(\omega) = \sum_k |g_k|^2 \delta(\omega - \omega_k), \tag{26}$$

how the coupling strengths are distributed across various frequencies. One important factor that establishes the characteristics of the bath and its impact on the system is the spectral density. Typical spectral densities consist of *Ohmic*

$$J(\omega) \propto \omega e^{-\frac{\omega}{\omega_c}} \tag{27}$$

(where the cutoff frequency is ω_c) symbolizes a bath with a wide variety of frequencies and the *super-Ohmic*

$$J(\omega) \propto \omega^s e^{-\frac{\omega}{\omega_c}} \tag{28}$$

(where $s > 1$) higher frequencies with a stronger coupling. Next, *sub-Ohmic*

$$J(\omega) \propto \omega^s \tag{29}$$

(where $0 < s < 1$) lower frequencies with a stronger coupling.

As discussed earlier, the spectral density determines the nature of the bath's influence on the system. **Figure 1** provides a visual representation of these spectral

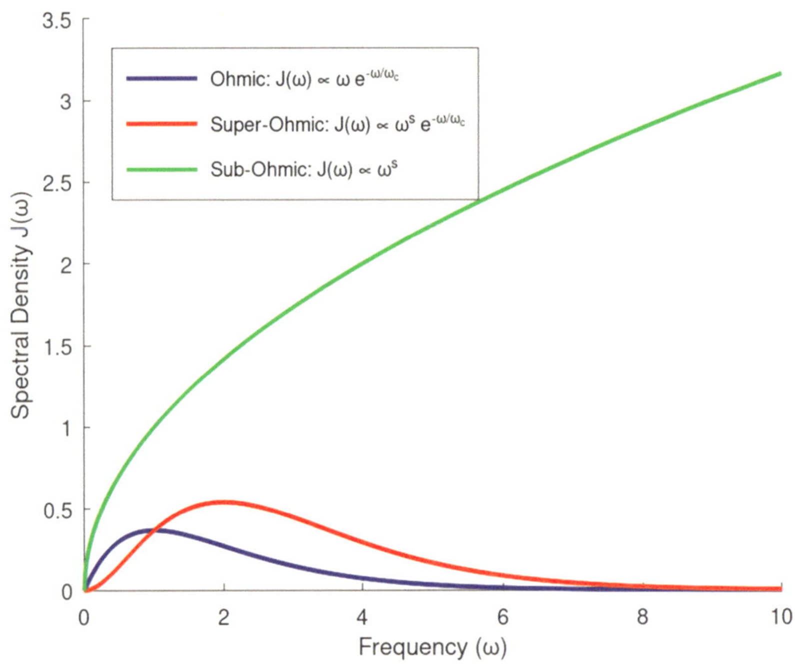

Figure 1.
Spectral densities for Ohmic, super-Ohmic, and sub-Ohmic baths. The plot illustrates the frequency dependence of the spectral density, $J(\omega)$, for each bath type. Parameters used the cutoff frequency $\omega_c = 1, s = 2$ for super-Ohmic, and $s = 0.5$ for sub-Ohmic.

densities for Ohmic, Super-Ohmic, and sub-Ohmic baths, illustrating the differences in their frequency dependence.

Figure 1 displays the spectral densities of three different types of baths: Ohmic, super-Ohmic, and sub-Ohmic, which are plotted as a function of frequency (ω). The x-axis represents the frequency, ranging from 0 to 10, while the y-axis represents the spectral density, $J(\omega)$. The Ohmic spectral density increases linearly with frequency up to a certain point, then decreases exponentially due to the exponential cutoff term. The super-Ohmic spectral density exhibits a stronger increase at higher frequencies compared to the Ohmic case, also with an exponential cutoff. The sub-Ohmic spectral density shows a slower increase with frequency, indicating a stronger coupling at lower frequencies. In nonequilibrium environments, these distinct spectral densities dictate how the system dissipates energy and evolves toward a steady state, influencing the rates of relaxation and decoherence. These spectral densities are critical for understanding the relaxation and decoherence processes that occur when a system is driven out of equilibrium and interacts with its environment. The specific form $J(\omega)$ determines the timescales and pathways for energy dissipation and the loss of coherence.

5. Quantifying coherence of quantum Otto engine on thermodynamic performance

Measuring coherence is essential to comprehending the thermodynamic performance of the quantum Otto engine. We can examine how coherence dynamics directly affect the engine's efficiency by creating coherence metrics through simulations. We can investigate the effects of coherence control techniques on improving performance under different nonequilibrium circumstances by establishing a relationship between the degree of coherence and the engine's thermodynamic results. This study sheds light on how efficiency can be maximized by preserving or adjusting coherence, especially in quantum systems that function far from equilibrium. These components can be combined to provide a thorough framework for assessing and enhancing quantum Otto engine performance, which will further the study of quantum thermodynamics.

5.1 Developing coherence measures

To quantify the amount of coherence present in the system at various stages of the thermodynamic cycle, which has two adiabatic and isothermal stages, we used the established measures of quantum coherence of the quantum Otto engine (qubit) for relative entropy of coherence and 1-norm of coherence. We created the coherence measures in the following manner using the density matrix principles: Density Matrix Diagonalization:

$$\rho_{\text{diag}} = \text{diag}(\text{diag}(\rho)) \tag{30}$$

We add a tiny positive value ε as part of the regularization process for numerical stability:

$$\varepsilon = 10^{-10} \tag{31}$$

Regularization is applied to the density matrix and its diagonal form:

$$\begin{aligned} \rho_{\text{regularized}} &= \rho + \varepsilon I \\ \rho_{\text{diag,regularized}} &= \rho_{\text{diag}} + \varepsilon I \end{aligned} \tag{32}$$

where I is the same-sized identity matrix as ρ.

The regularized density matrices' logarithm is calculated as follows:

$$\log\left(\rho_{\text{regularized}}\right) \quad \text{and} \quad \log\left(\rho_{\text{diag,regularized}}\right) \tag{33}$$

Relative entropy of coherence calculation: Numerical stability is checked:

$$\text{if any}\left(\text{NaN}\left(\log\left(\rho_{\text{regularized}}\right)\right)\vee\text{any}\left(\inf\left(\log\left(\rho_{\text{regularized}}\right)\right)\right)\right) \tag{34}$$

We assign

$$C_{\text{rel}} = \infty, \tag{35}$$

if this condition is satisfied. Otherwise, the following formula is used to calculate the relative entropy of coherence:

$$C_{\text{rel}} = \text{tr}\left(\rho_{\text{regularized}}\left(\log\left(\rho_{\text{regularized}}\right) - \log\left(\rho_{\text{diag,regularized}}\right)\right)\right) \tag{36}$$

5.2 Analyzing the dynamics of coherence

To learn how coherence is created, preserved, and destroyed during the isothermal and thermostatic processes, it is necessary to monitor the changes in coherence measurements over time. Coherence function for the isothermal process:

$$C_{isothermal}(t) = C_0 e^{-t} \tag{37}$$

where (C_0) represents the initial measure of coherence at time $(t = 0)$. This equation illustrates how the coherence of the quantum state diminishes exponentially over time due to decoherence effects. Specifically, the factor (e^{-t}) indicates that as time progresses, the coherence decays, resulting in a reduction of quantum superposition and entanglement within the system, which is characteristic of isothermal processes where temperature remains constant. In contrast, the coherence function for the thermostatic process is described by the equation

$$C_{thermostatic}(t) = C_0\left(1 - e^{-5t}\right) \tag{38}$$

This formulation indicates that coherence increases over time instead of decaying. The initial coherence measure (C_0) is multiplied by the term $((1 - e^{-5t}))$, which reflects that coherence approaches its maximum value as time increases. The factor $(5t)$ within the exponential indicates a rapid buildup of coherence compared to the isothermal case, suggesting that the thermostatic process favors the maintenance or enhancement of coherence over time. This behavior may be associated with the interactions that stabilize the quantum system, facilitating sustained coherence, unlike in the isothermal scenario (**Figure 2**).

The dynamics of quantum coherence during the isothermal and thermostatic phases, two crucial steps in the quantum Otto cycle, are depicted in this figure.

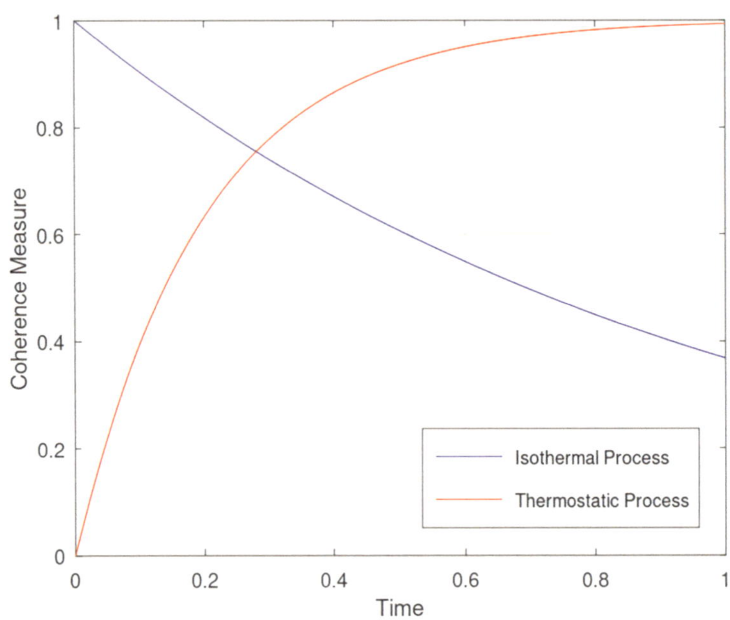

Figure 2.
Coherence in quantum Otto engine processes.

Coherence decays exponentially in the isothermal process. This is in line with scientific knowledge that when a system undergoes isothermal expansion or compression, it interacts strongly with a thermal bath, causing decoherence as the system's quantum state entangles with the many degrees of freedom of the bath. The rate of decay reflects both the bath's temperature and the strength of the system-bath coupling. On the other hand, the thermostatic process exhibits a gradual increase in coherence. This implies that the system may be pushed in the direction of a more coherent state under particular circumstances, perhaps by means of artificial interactions or outside forces that oppose decoherence. Two crucial factors that affect the engine's performance are the rate of increase and the saturation level of coherence.

5.3 Correlating coherence with thermodynamic performance

Examining the connection between coherence metrics and important engine performance indicators like entropy production, efficiency, and work output [11].

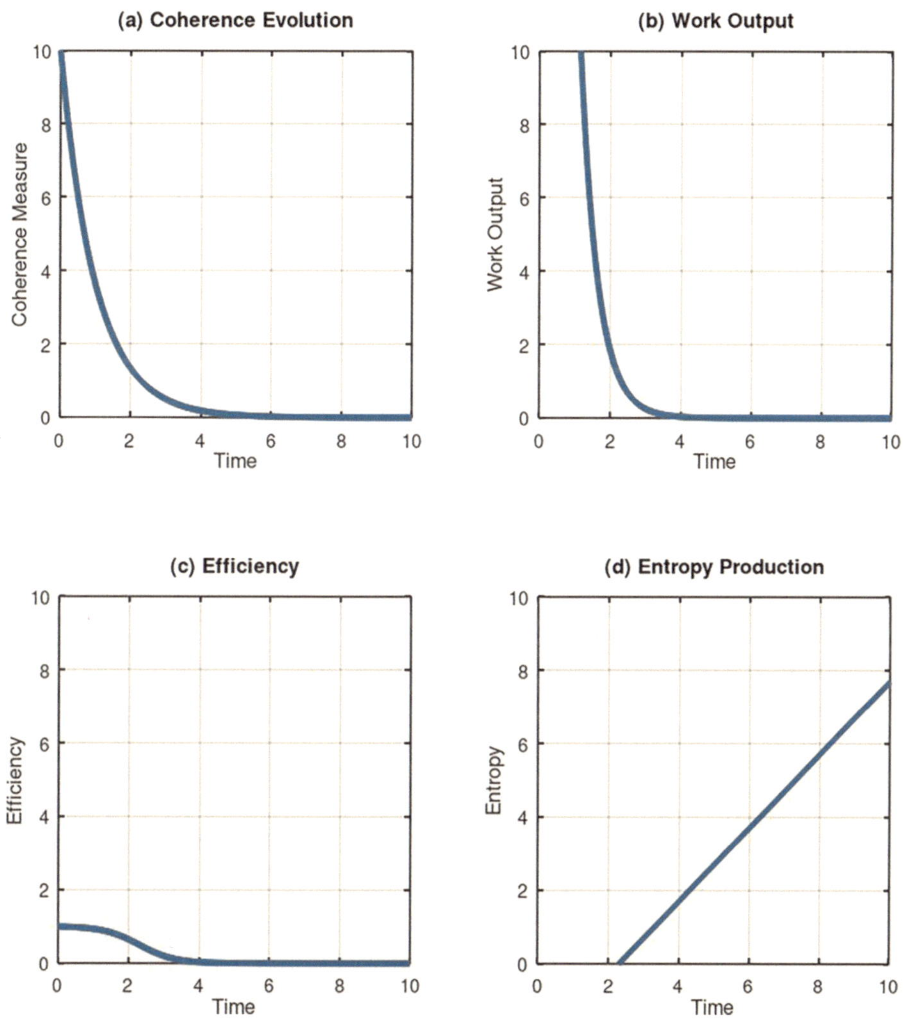

Figure 3.
Performance metrics of the quantum Otto engine.

Figure 3 links coherence to important thermodynamic metrics to provide a comprehensive picture of the engine's performance. A measure of coherence within the working substance that varies with time is probably depicted in the coherence evolution **Figure 3 (a)**. The engine's capacity to draw energy from the thermal reservoirs is shown by the work output in **Figure 3(b)**. For the cycle to be optimized, the curve's shape—particularly its peak value and duration—is essential. The engine's ability to convert heat into work is measured by the efficiency, as shown in **Figure 3(c)**. Less energy wasted as heat is implied by higher efficiency. Lastly, the irreversibility of the cycle is shown in the entropy production in **Figure 3(d)**. Because it suggests a more reversible and effective process, lower entropy production is preferred. It is possible to deduce how coherence directly affects the engine's thermodynamic performance by examining these plots collectively.

5.4 Exploring coherence control strategies

Establishing methods to adjust and regulate coherence in the system, such as by applying external driving fields or modifying the system-bath coupling, in order to maximize engine performance [29]. Coherence model with coupling and external driving field effects:

$$C(t) = C_0 e^{-\gamma t} + \sin 2\pi t \tag{39}$$

The time-dependent coherence (C(t)) of the quantum system, which is essential for the effectiveness of the quantum Otto engine [8], is modeled by this equation, where C_0 is the initial coherence measure and γ is the coupling strength. The term $(C_0 e^{-\gamma t})$ denotes the decay of initial coherence due to interactions with the environment (influenced by the coupling strength (γ)). According to McKeever et al. [29], as time passes, coherence decreases, which impacts the system's capacity to utilize quantum mechanical advantages. The effects of external driving fields are taken into account by the second term, $(\sin 2\pi t)$, which contributes periodic oscillations that can momentarily improve coherence. In order to achieve optimal work output in the cycle, this model emphasizes the delicate balance between coherence loss and potential revival through external influences [30]. Work output in relation to coherent interactions with the environment (affected by the coupling strength (γ)). Coherence decreases with time, which impacts the system's capacity to utilize the benefits of quantum mechanics. $W = C^2$ According to this equation, the quantum Otto engine's work output (W) is a quadratic function of coherence (C). It shows that the work generated is directly proportional to the square of the coherence level; therefore, higher coherence leads to significantly greater work extraction [31, 32]. This relationship highlights the significance of preserving high coherence in the engine's states because it increases the capacity to perform useful work, which is consistent with quantum thermodynamic efficiency, where coherence improves the energy conversion process. Efficiency is based on work output $\eta = \frac{W}{1+W}$. In this context, efficiency (η) is the relationship between the total energy input, represented as (1 + W), and the work output (W). According to this formulation, the system's effective energy is equal to the work performed plus a baseline energy (normalized to 1). Realistic restrictions on energy conversion processes are reflected in the efficiency [33], which rises as work output rises and approaches but never quite reaches the maximum value of 1. This efficiency formula shows how the quantum Otto engine's overall performance can be enhanced by increased coherence and work output. The relationship between coherence and entropy production: $S = -\log(C + 1 \times 10^{-6})$ (where 1×10^{-6} is added to avoid $\log(0)$). The entropy production (S) with respect to coherence (C) is defined by this equation. The notion that

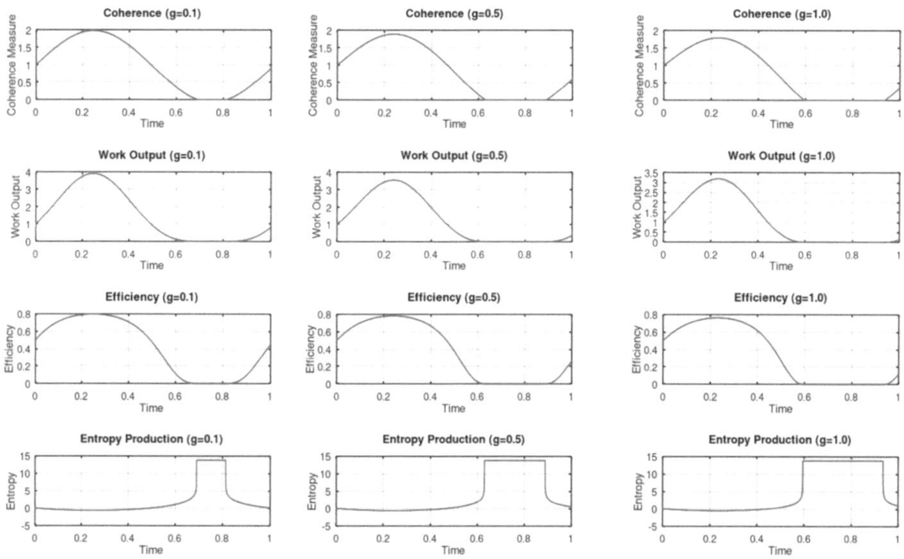

Figure 4.
Quantum Otto engine coherence control simulation.

entropy decreases with increasing coherence, suggesting a more ordered quantum state [34], is reflected in the negative logarithm of coherence (with a tiny offset $\left(1 \times 10^{-6}\right)$ to prevent mathematical singularities). This relationship illustrates how preserving coherence can lower entropy and improve performance in the context of the quantum Otto engine. According to Gonzalo Manzano et al. [35], the laws of quantum thermodynamics state that a system with lower entropy production has a greater capacity to extract work effectively from thermal reservoirs (**Figure 4**).

The effect of coupling strength (γ) on engine performance is examined in this figure. The relationship between the system and its surroundings is represented by the coupling strength. The figure probably illustrates how changing γ results in distinct coherence dynamics, which in turn impact efficiency, entropy production, and work output. While a strong coupling (large γ) may accelerate decoherence while enhancing energy exchange, a weak coupling (small γ) may limit energy exchange while maintaining coherence. A trade-off between these conflicting effects is represented by the ideal γ value. The data in the figure can be used to determine the optimal coupling strength [36] for enhancing engine performance in particular operating scenarios.

5.5 Coherence and efficiency

Coherence can improve quantum engine performance beyond classical bounds, according to earlier research. Coherence, for example, may enhance work output and allow for greater resilience against thermal noise [5, 37]. The given equations can be obtained from a general framework that incorporates the ideas of performance in energy conversion processes and coherence in quantum systems [38, 39].

5.5.1 Coherence decay model

A quantum system's decaying coherence over time can be represented as $C(t) = C_0 e^{-\lambda t}$ where the decay constant is represented by λ. The coherence decay in a quantum system over time is represented by this equation. Where $C(t)$ represents the coherence at time (t), C_0 is the initial coherence value, which represents the maximum coherence when the system is perfectly coherent at $(t = 0)$, and λ is the decay constant, which indicates the rate at which the coherence decreases with time. A loss of information and performance in quantum tasks is indicated by the exponential decline in the coherence of the quantum state over time.

5.5.2 Quantum performance with coherence

In coherent quantum systems, the ability to maintain these superpositions enables more effective quantum interference, allowing for optimized energy transfer and reduced decoherence impacts. This can lead to improved performance in quantum technologies, such as quantum computing and quantum thermodynamics, where maintaining coherence is essential for achieving tasks like information processing and energy conversion. Consequently, the interplay between coherence, efficiency, and work output highlights the importance of quantum coherence in realizing the full potential of quantum mechanical systems in practical applications. Coherence in a quantum system can be linked to efficiency and work output:

$$W_{quantum} = f(C)^2 = C(t)^2 = \left(C_0 e^{-\lambda t}\right)^2$$

$$\eta_{quantum} = \frac{W_{quantum}}{1 + W_{quantum}} = \frac{C(t)^2}{1 + C(t)^2}$$

The square of a quantum system's coherence determines how much work it produces. This formula uses a quantum system's work output to determine its efficiency. Because of the quadratic relationship with work output, efficiency rises with coherence [40, 41]. Efficiency increases with high coherence, highlighting the significance of preserving quantum states for peak performance.

5.5.3 Classical performance model (without coherence)

In quantum Otto engines, thermal noise plays a critical role in determining the work output by introducing fluctuations that affect the system's efficiency. As these engines operate between two thermal reservoirs, the random energy exchanges due to thermal noise can lead to variations in the engine's performance during the adiabatic and isothermal processes. High levels of thermal noise can result in energy losses that hinder the engine's ability to convert heat into work effectively. Therefore, understanding and mitigating the effects of thermal noise are essential for optimizing the work output of quantum Otto engines, allowing for improved performance in converting thermal energy into mechanical work. Thermal noise determines the work output in a classical system.

$$W_{classical} = g(\text{thermal\_noise}) = 1 - \text{thermal\_noise}$$

$$\eta_{classical} = W_{classical}$$

The amount of energy that can be used, less the energy lost to thermal noise, is the work output in classical systems. The direct effect of thermal noise on energy conversion is reflected in the fact that the efficiency of a classical system is simply equal to its work output. Since coherence is not a factor in efficiency in classical systems, thermal noise limits performance, leading to a simpler relationship. The combined framework could put these components together in a single equation that encapsulates both quantum and classical performance under thermal noise and coherence, $W = W_{quantum} + W_{classical}$, and $\eta = \frac{W}{1+W}$ where η is the overall efficiency and W is the total work output taking into account both quantum coherence decay and classical performance. Using this framework, you can examine various quantum and classical system regimes while accounting for the effects of thermal noise and coherence. Coherence is emphasized in the quantum model, which demonstrates how preserving coherence can result in improved efficiency and performance. In contrast, efficiency is directly correlated with usable energy in the classical model, which is simple in terms of energy loss from thermal effects.

A direct comparison between a classical Otto engine without coherence and a quantum Otto engine with coherence is shown in this figure. Utilizing quantum coherence is advantageous, as evidenced by the quantum engine's increased efficiency and work output. This "quantum advantage" of using coherence is quantified by the difference between the two curves. This illustration demonstrates how quantum thermodynamics has the ability to convert energy beyond classical bounds.

While classical systems are limited by thermal fluctuations, quantum systems can use coherence to achieve better performance, as explained by the extracted equations. To model and optimize the systems being studied, you can change parameters such as C_0, λ, and thermal noise. Each of these equations illustrates how quantum mechanical concepts relate to thermodynamics and clarifies the functions of coherence, energy transitions, and overall efficiencies in the functioning of the quantum Otto engine.

5.6 Contribution of coherence on quantum Otto engine

Employing MATLAB/Octave, a program was created to model a quantum Otto cycle and examine how various coherence levels affect productivity and efficiency. Representation of density matrix: The quantum Otto cycle's initial density matrix ρ is defined as

$$\rho_{\text{initial}} = \text{coherence\_init}^2 \begin{bmatrix} 1 & 0 \\ 0 & 0 \end{bmatrix} + \left(1 - \text{coherence\_init}^2\right) \begin{bmatrix} 0 & 0 \\ 0 & 1 \end{bmatrix}$$

The initial density matrix (ρ) that depicts the system's quantum state is defined by this equation. Two pure states are combined in the density matrix: the excited state (with the complementary probability) and the ground state (with coherence squared, which represents the probability of being in that state). Each state's contribution is altered by the parameter (coherence_init), where a higher coherence denotes a stronger presence of one state relative to the other. The density matrix and reservoir temperature are used to compute the energy from the thermal states:

$$E_{\text{hot}} = -\log\left(\text{trace}\left(\exp\left(-\frac{\rho_{\text{initial}}}{T_{\text{hot}}}\right)\right)\right) \tag{40}$$

$$E_{\text{cold}} = -\log\left(\text{trace}\left(\exp\left(-\frac{\rho_{\text{current}}}{T_{\text{cold}}}\right)\right)\right) \tag{41}$$

The system's energy in thermal equilibrium with the heated reservoir is calculated using this equation. The temperature-scaled trace of the exponential of the negative density matrix is used. In order to determine how much energy is available for work extraction, the free energy associated with the quantum state is evaluated using the logarithm. This formula computes the energy when the system interacts with the cold reservoir, just like it does with the hot reservoir. The modified density matrix after taking coherence decay into account is denoted by (ρ_{current}). The difference in energy between the hot and cold reservoirs is used to compute the amount of work completed at each stage of the simulation:

$$W_{\text{done}} = E_{\text{hot}} - E_{\text{cold}} \tag{42}$$

The difference between the energies determined for the hot and cold reservoirs is the amount of work completed at each stage of the cycle. Positive work is extracted if the hot reservoir's energy is higher; if not, the system may absorb energy. The sum of the work produced during the simulation steps is calculated:

$$\text{work\_output+} = W_{\text{done}} \tag{43}$$

By adding the work completed in each iteration, this shows the total amount of work produced during all simulation steps. It shows how much energy was taken out overall during the Otto cycle. Following each step, the coherence exponentially decays:

$$\text{current\_coherence} = \text{coherence\_init} \cdot e^{-\text{decay\_rate}\cdot\text{step}} \tag{44}$$

This equation represents the decay of coherence over time, with *decay_rate* dictating the rate of decay. Coherence falls off exponentially with each simulation step, simulating the loss of distinguishability in quantum states. Using the total work output divided by the total energy supplied by the hot reservoir, the cycle's efficiency can be computed:

$$\text{efficiency} = \frac{\text{work\_outputs}}{T_{\text{hot}} \cdot \text{num\_steps}} \tag{45}$$

The ratio of total work output to total energy supplied by the hot reservoir over all steps is the definition of efficiency in this context. It gauges how well the system uses the quantum Otto cycle to transform thermal energy into work. To make the work outputs comparable, normalize them:

$$\text{normalized\_work\_outputs} = \frac{\text{work\_outputs}}{\max(\text{work\_outputs})} \tag{46}$$

The work outputs are rescaled during this normalization process to fall between 0 and 1. By acting as a benchmark, the maximum work output makes it easier to compare various coherence levels without losing sight of their relative magnitudes.

The relationship between efficiency and coherence level is plotted explicitly in **Figure 5**. The increasing trend implies that efficiency increases with coherence. The nature of this relationship is revealed by the curve's shape (linear, quadratic, etc.). Further context is given by the plots of total work output and normalized total work output, which demonstrate the impact of coherence on the overall energy extraction. To comprehend the quantitative advantages of keeping the engine coherent, this figure is crucial (**Figure 6**).

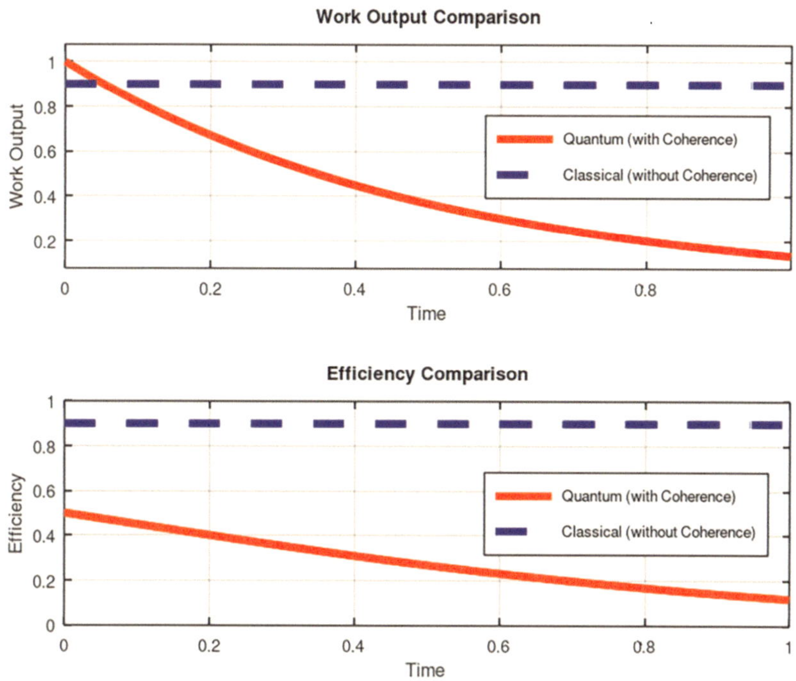

Figure 5.
Coherence enhancement in quantum Otto engine performance.

6. Nonequilibrium conditions in quantum systems

Nonequilibrium conditions occur when a quantum system fails to achieve thermal equilibrium with its surroundings; in the case of quantum Otto engines, these conditions may include time-dependent Hamiltonians, fluctuating coupling strengths, or different temperatures. Temperature profiles for the hot and cold reservoirs:

$$T_{hot}(t) = 1 + 0.5\sin(2\pi t) \quad (47)$$

This equation models the temperature of a hot reservoir that fluctuates sinusoidally over time, oscillating between 0.5 and 1 units. This reflects a periodic variation in temperature, common in nonequilibrium systems where energy inputs or extraction vary cyclically.

$$T_{cold}(t) = 0.5 + 0.2\cos(2\pi t) \quad (48)$$

Similarly, this equation describes the temperature of a cold reservoir that changes over time, oscillating between 0.2 and 0.5 units. The cosine fluctuation suggests that the cold reservoir experiences periodic cooling and heating, contributing to the nonequilibrium state. Fluctuating coupling strength:

$$g(t) = 0.1 + 0.9\sin(2\pi t) \quad (49)$$

This equation represents the strength of the coupling between the hot and cold reservoirs, varying periodically between 0.1 and 0.9. Such fluctuations are significant

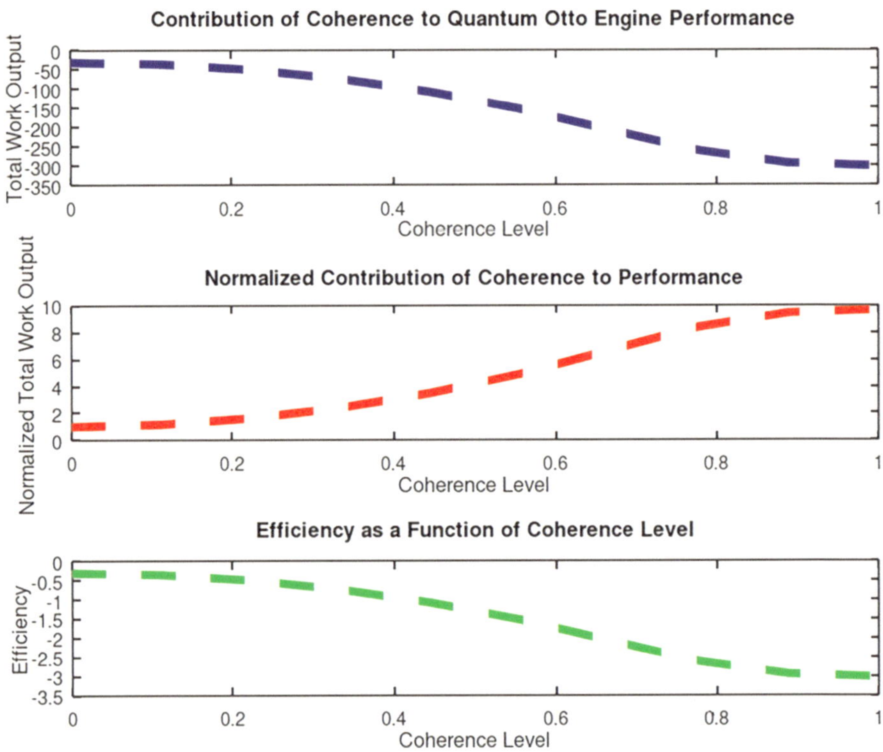

Figure 6.
Simulation of the performance of a quantum Otto engine over 100 steps, illustrating the impact of varying coherence levels on work output. The engine operates between a hot reservoir at a temperature of 1.0 and a cold reservoir at 0.1, while accounting for a coherence decay rate of 0.01 over time.

in nonequilibrium environments because they affect the transfer of energy between the two reservoirs, influencing the overall system dynamics. Time-dependent Hamiltonian:

$$H(t) = 1 + 0.5\cos(4\pi t) \tag{50}$$

The Hamiltonian describes the total energy of the system, which also varies over time due to the cosine oscillation. This indicates that the energy landscape of the system is changing dynamically, which is typical in nonequilibrium scenarios where external conditions or interactions are not fixed. Temperature difference:

$$\Delta T = T_h - T_c \tag{51}$$

This equation computes the temperature difference between the hot and cold reservoirs. In nonequilibrium environments, a positive difference (when $(T_h > T_c)$) is essential for driving energy transfer processes, as this gradient is the force behind heat flow. Work output formula:

$$\text{curr\_work} = \begin{cases} g \cdot \Delta T \cdot e^{-H}, & \text{if } \Delta T > 0 \\ 0, & \text{otherwise} \end{cases} \tag{52}$$

This equation calculates the work output of the system based on the coupling strength, temperature difference, and the exponential decay factor related to the

Hamiltonian. It highlights the efficiency of energy conversion processes in nonequilibrium environments, indicating that work is produced only when there is a temperature gradient. The exponential term suggests that as energy in the system increases (as controlled by H), the work output decreases, which might represent losses due to nonconservative processes. Efficiency calculation:

$$\text{curr\_efficiency} = \frac{\text{curr\_work}}{T_h + 1 \times 10^{-6}} \tag{53}$$

This equation defines the efficiency of the work output relative to the temperature of the hot reservoir. The addition of a small term (1×10^{-6}) ensures that the denominator does not approach zero, which is crucial for maintaining defined efficiency calculations even at extremely low temperatures. In nonequilibrium settings, this efficiency metric is critical for assessing how effectively the temperature gradient is harnessed for work (**Figure 7**).

The amplitude and frequency of these oscillations are crucial parameters that characterize the engine's response to nonequilibrium conditions. This figure exposes the difficulties of running quantum engines in realistic environments and the necessity for strong control strategies. The work output and efficiency plots most likely exhibit oscillations or fluctuations, reflecting the time-dependent nature of the environment (e.g., fluctuating temperatures).

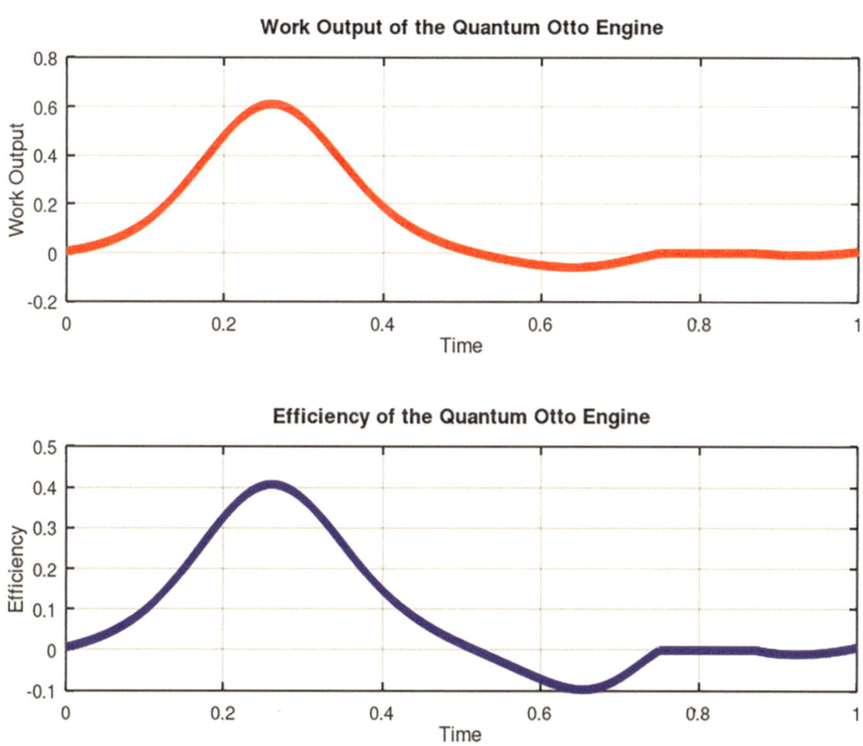

Figure 7.
Simulation of quantum Otto engine in nonequilibrium environments.

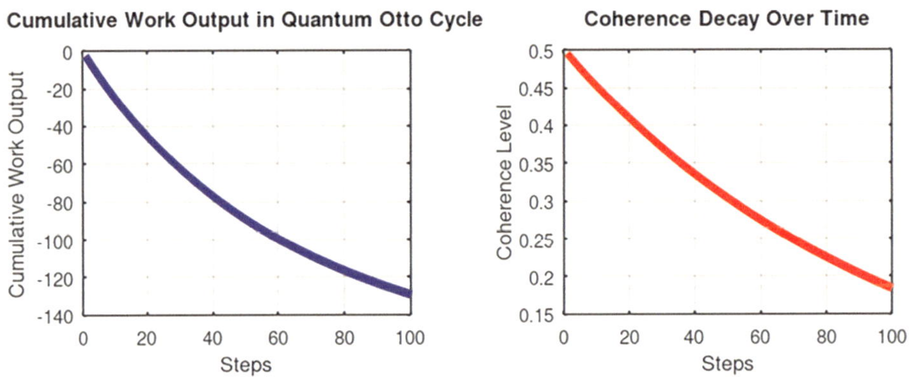

Figure 8.
Simulation of the quantum Otto cycle under varying levels of coherence and nonequilibrium.

6.1 Numerical simulations

Models that simulate the quantum Otto cycle under various coherence and nonequilibrium parameters were implemented using MATLAB/Octave. We set the number of steps to 100, the number of time steps for the simulation temperature in a hot bath to 1.0, the temperature of the hot reservoir in a cold bath to 0.1, the temperature of the cold reservoir's initial coherence to 0.5, and the initial coherence level to 1.0. The results indicate a final coherence level of 0.1839 and a total work output of −129.3280 (**Figure 8**).

The effects of different coherence levels and nonequilibrium conditions on the engine's performance are combined in this figure. The coherence decay plot illustrates how coherence deteriorates over time, while the cumulative work output plot displays the total energy extracted over several cycles. The engine's long-term stability and efficiency under realistic operating conditions can be evaluated by combining these plots [42]. This number is essential for creating workable quantum heat engines that can continue to operate well even in the presence of noise from the environment.

6.2 Analytical techniques

Utilizing both perturbative and nonperturbative methods to gain an understanding of how coherence improves performance in the presence of environmental interactions (**Figure 9**).

The figure compares the two methods. The system-environment interactions are strong, and a perturbative treatment is insufficient if the curves from the perturbative and nonperturbative methods differ significantly [17]. The nonperturbative method offers a more accurate assessment of the engine's efficiency and the function of coherence in these situations. The figure offers insights into the circumstances in which coherence can be successfully harnessed for increased efficiency and aids in validating the theoretical models used to describe the quantum Otto engine.

7. Practical implications

Investigating quantum coherence under nonequilibrium conditions is a novel field of study with important applications in many different fields of science and

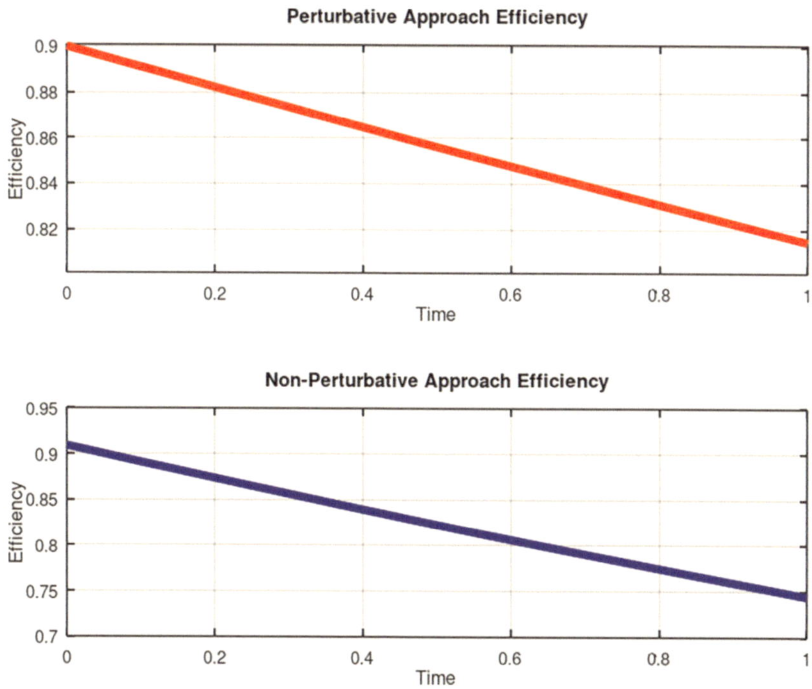

Figure 9.
Perturbative and nonperturbative approaches in deriving how coherence enhances efficiency amidst environmental interactions.

technology. The potential for revolutionary applications, ranging from quantum computing to energy efficiency and advanced material science, becomes more evident as more research reveals the complex interactions between these theoretical constructs.

7.1 Quantum computing and information processing

The development of quantum computing technologies is one of the most important consequences of comprehending the connection between quantum coherence and nonequilibrium states. The superposition states that allow qubits to perform computations far more efficiently than classical bits depend on quantum coherence. Coherence is crucial for obtaining reduced error rates in quantum gates, which are essential parts of quantum circuits, according to research by [43]. Coherence under nonequilibrium conditions makes it feasible to create more reliable quantum algorithms that can execute intricate computations more quickly, opening the door for advances in cryptography and optimization issues.

7.2 Energy-efficient devices

Another crucial use of this research is the creation of effective quantum heat engines. The subtleties of quantum systems functioning in nonequilibrium conditions, which can be used to greatly improve performance, are not taken into consideration by traditional thermodynamics. According to Kosloff and Levy [44], these engines can perform better than their classical counterparts in terms of work output per cycle because they are based on concepts that take advantage of coherence. Innovations in sustainable energy technologies may result from this, advancing greener energy solutions that reduce waste and boost energy conversion processes' overall effectiveness.

7.3 Advancements in materials science

Understanding quantum coherence in nonequilibrium settings can help materials scientists create new materials with remarkable qualities. Maintaining coherence in materials can lead to improved conductivity or special optical properties, both of which are critical for photonics and electronics applications. For instance, studies on superconducting qubits by Nakamura et al. [45] show how important coherence is to achieving effective energy transfer, pointing the way toward the creation of materials that function well at room temperature.

7.4 Quantum networks and communication

The implications also apply to quantum communication systems, where quantum networks' performance and dependability can be improved by preserving coherence in nonequilibrium states. The potential for more secure communication protocols using quantum entanglement and coherence is highlighted by research by Gisin et al. [46]. Effective coherence management and utilization will be critical as quantum networks develop in order to create reliable connections over greater distances, which is critical for the developing quantum internet.

7.5 Advanced sensing technologies

Furthermore, revolutionary improvements in sensing technologies are possible through an understanding of quantum coherence under nonequilibrium conditions. By taking advantage of quantum superposition and entanglement, quantum sensors are able to achieve sensitivity levels that are significantly higher than those of classical sensors. Research by Van der Wal et al. [47] shows how quantum sensors can identify even the smallest changes in environmental parameters, leading to innovations in navigation, environmental monitoring, and medical diagnostics.

7.6 Energy storage innovations

Finally, novel approaches to energy storage could result from the interplay between quantum coherence and nonequilibrium states. Advanced batteries and supercapacitors with improved storage capacities and quicker charge-discharge cycles can be designed by optimizing quantum states for energy capture and release. Batteries with considerably shorter charging times and longer lifespans could result from research into these systems, which would help power grids incorporate renewable energy sources.

An extensive range of real-world applications in various industries is made possible by the study of quantum coherence and nonequilibrium conditions. The potential to convert these theoretical insights into practical innovations grows more promising as research continues to expand our understanding. A more connected and sustainable technological future can be achieved by embracing the concepts of quantum mechanics to address current issues in computing, energy efficiency, materials development, and other fields. Future engine designs that are optimized for quantum computation, renewable energy sources, or nanoscale devices may benefit from these insights. Additionally, developing more accurate thermodynamic frameworks that take coherence effects into account may aid in the development of new protocols for quantum systems.

8. Conclusion

To sum up, this chapter has investigated the complex connection between nonequilibrium quantum Otto engine performance and quantum coherence. We have demonstrated how preserving quantum coherence can greatly improve engine performance, resulting in higher work extraction and efficiencies when compared to classical counterparts, by utilizing the Quantum Master Equation framework and creating particular coherence measures.

The difficulties and possibilities of using coherence for useful purposes were demonstrated by our research into nonequilibrium circumstances, such as changing temperatures and coupling strengths. The analytical methods and simulations that are offered provide insightful information about how to optimize engine parameters for reliable performance in realistic settings.

Quantum computing, energy-efficient devices, materials science, quantum networks, and advanced sensing technologies are just a few of the domains where this research has practical implications. We can open the door for revolutionary inventions that take advantage of the special qualities of quantum systems by comprehending and adjusting coherence.

With a focus on the potential of coherent quantum states to power next-generation energy systems and quantum technologies, this work connects theoretical developments in quantum thermodynamics with real-world applications. To further improve coherence and realize the full potential of quantum Otto engines, future research should concentrate on creating increasingly complex control schemes and investigating new materials.

Acknowledgements

The authors would like to acknowledge the support and resources that made this research possible. We thank each one of the authors for providing the necessary conducive environment for conducting this study in their tight schedules.

AI disclosure statement

The authors acknowledge the use of AI for language polishing of the manuscript.

This manuscript (Paper) has undergone the review and editing process using artificial intelligence tools QuillBot and Grammarly. These tools have been utilized to enhance grammar, improve sentence structure, and ensure clarity and coherence throughout the text.

QuillBot is employed to paraphrase and refine wording for better expression, while Grammarly assists in identifying grammatical errors and suggesting corrections for optimal readability. While these AI tools contribute significantly to the writing process, the final responsibility for the content and its accuracy remains with the authors.

Author details

Jeolous Malamula Nyasulu[1,2*], Chunyang Wang[1], Zengxuan Zhao[1], Mushtaq Rana Imran[1,3], Elson Rodger Katete[4], Akuonani Zakeyo Phiri[2] and Evans Kumalo Mtambo[2,4]

1 Institute of Theoretical Physics, School of Physics and Optoelectronic Engineering, Ludong University, Yantai, China

2 Department of Physics, Domasi College of Education, Domasi, Zomba, Malawi

3 School of Science, Harbin Institute of Technology, Shenzhen, China

4 Department of Physics and Electronics, Mzuzu University, Mzuzu, Malawi

*Address all correspondence to: jnyasulu688@gmail.com

References

[1] Abah O, Roßnagel J, Jacob G, Deffner S, Schmidt-Kaler F, Singer K, et al. Single-ion heat engine at maximum power. Physical Review Letters. 2012; **109**:203006

[2] Kedem O, Entin-Wohlman O. Thermodynamic performance of a quantum otto engine under non-equilibrium conditions. New Journal of Physics. 2019;**21**(7):073024

[3] Entin-Wohlman O, Imry Y, Aharony A. Enhanced performance of joint cooling and energy production. arXiv preprint arXiv:1410.4880. 2014

[4] Kosloff R, Rezek Y. The quantum harmonic Otto cycle. Entropy. 2017;**19** (4):136

[5] Scully MO, Chapin KR, Dorfman KE, Kim MB, Svidzinsky A. Quantum heat engine power can be increased by noise-induced coherence. Proceedings of the National Academy of Sciences. 2011;**108** (37):15097-15100

[6] Francica G, Goold J, Plastina F. Role of coherence in the nonequilibrium thermodynamics of quantum systems. Physical Review E. 2019;**99**(4):042105

[7] Esposito M, Lindenberg K, Van den Broeck C. Universality of efficiency at maximum power. Physical Review Letters. 2009;**102**(13):130602

[8] Özdemir AT, Müstecaplioğlu ÖE. Quantum thermodynamics and quantum coherence engines. Turkish Journal of Physics. 2020;**44**(5):404-436

[9] Xu G-H, Jiang C, Minami Y, Watanabe G. Relation between fluctuations and efficiency at maximum power for small heat engines. Physical Review Research. 2022;**4**(4):043139

[10] Zhang Y, Liao J. Quantum otto heat engines with interacting particles. Physical Review A. 2018;**98**(3):032110

[11] Xiao Y, Liu D, He J, Ma Y, Zhaoqi W, Wang J. Quantum Otto engine with quantum correlations. Physical Review A. 2023;**108**(4):042614

[12] Hewgill A, Ferraro A, De Chiara G. Quantum correlations and thermodynamic performances of two-qubit engines with local and common baths. Physical Review A. 2018;**98**(4): 042102

[13] Perarnau-Llobet M, Hovhannisyan KV. Quantum heat engines: A perspective on working substances. Entropy. 2015;**17**(11):7750-7768

[14] Perarnau-Llobet M, Hovhannisyan KV, Huber M, Skrzypczyk P, Brunner N, Acín A. Extractable work from correlations. Physical Review X. 2015;**5** (4):041011

[15] Solfanelli A, Falsetti M, Campisi M. Nonadiabatic single-qubit quantum otto engine. Physical Review B. 2020;**101**(5): 054513

[16] Camati PA, Santos JFG, Serra RM. Coherence effects in the performance of the quantum otto heat engine. Physical Review A. 2019;**99**:062103

[17] Jacquod P, Petitjean C. Decoherence, entanglement and irreversibility in quantum dynamical systems with few degrees of freedom. Advances in Physics. 2009;**58**(2):67-196

[18] Koyanagi S, Tanimura Y. Classical and quantum thermodynamics in a non-equilibrium regime: Application to thermostatic stirling engine. arXiv preprint arXiv:2405.17791. 2024

[19] Pezzutto M, Paternostro M, Omar Y. An out-of-equilibrium non-markovian quantum heat engine. Quantum Science and Technology. 2019;**4**(2):025002

[20] Campisi M, Pekola J, Fazio R. Nonequilibrium fluctuations in quantum heat engines: Theory, example, and possible solid state experiments. New Journal of Physics. 2015;**17**(3):035012

[21] Rodrigues FLS, Lutz E. Nonequilibrium thermodynamics of quantum coherence beyond linear response. Communications Physics. 2024;**7**(1):61

[22] Kaur K, Rebari S, Singh V. Performance analysis of quantum harmonic otto engine and refrigerator under a trade-off figure of merit. Journal of Non-Equilibrium Thermodynamics. 2025;**50**(1):1-19

[23] Abah O, Lutz E. Efficiency of heat engines coupled to nonequilibrium reservoirs. EPL (Europhysics Letters). 2014;**106**(2):20001

[24] Kosloff R. Quantum thermodynamics: A dynamical viewpoint. Entropy. 2013;**15**(6):2100-2128

[25] Liu Z-Y, Xia Y-J, Man Z-X. Entropy production of a quantum system in non-equilibrium environment: The effect of coherence. Physica A: Statistical Mechanics and its Applications. 2025; **664**:130445

[26] Kołodyński J, Brask JB, Perarnau-Llobet M, Bylicka B. Adding dynamical generators in quantum master equations. Physical Review A. 2018;**98** (2):022115

[27] Guo B-q, Zhou N-n, Yang C-p, Yu C-s. Multifunctional quantum thermal device with initial-state dependence. Physica E: Low-dimensional Systems and Nanostructures. 2022;**134**(114930):1-5

[28] Gelbwaser-Klimovsky D, Niedenzu W, Kurizki G. Thermodynamics of quantum systems under dynamical control. Advances in Atomic, Molecular, and Optical Physics. 2015;**64**: 329-407

[29] McKeever T, Diba O, Nazir A. A finite-time quantum otto engine subject to control noise and enhancement techniques. arXiv preprint arXiv: 2412.15196. 2024

[30] Nautiyal VV, Watson RS, Kheruntsyan KV. A finite-time quantum Otto engine with tunnel coupled one-dimensional Bose gases. New Journal of Physics. 2024;**26**(6):063033

[31] Xu B-M, Zou J, Guo L-S, Kong X-M. Effects of quantum coherence on work statistics. Physical Review A. 2018;**97**(5): 052122

[32] Korzekwa K, Lostaglio M, Oppenheim J, Jennings D. The extraction of work from quantum coherence. New Journal of Physics. 2016;**18**(2):023045

[33] Paoli L, Cullen J. Technical limits for energy conversion efficiency. Energy. 2020;**192**:116228

[34] Manzano G, Zambrini R. Quantum thermodynamics under continuous monitoring: A general framework. AVS Quantum Science. 2022;**4**(2)

[35] Manzano G, Galve F, Zambrini R, Parrondo JMR. Entropy production and thermodynamic power of the squeezed thermal reservoir. Physical Review E. 2016;**93**(5):052120

[36] Sarmah MJ, Bansal A, Goswami HP. Nonequilibrium fluctuations in boson transport through squeezed reservoirs.

Physica A: Statistical Mechanics and its Applications. 2023

[37] Tirone S, Salvia R, Chessa S, Giovannetti V. Quantum work extraction efficiency for noisy quantum batteries: The role of coherence. Physical Review A. 2025;**111**(1):012204

[38] Goold J, Hovhannisyan KV, Nourmandipour A, Perarnau-Llobet M. From quantum to classical thermodynamics: The otto cycle. Nature Communications. 2016;7:12493

[39] Binder F, Vinjanampathy S, Modi K, Goold J. Quantum thermodynamics of general quantum processes. Physical Review E. 2015;**91**(3):032119

[40] Deutsch IH. Harnessing the power of the second quantum revolution. PRX Quantum. 2020;**1**(2):020101

[41] Martin L. Quantum Feedback for Measurement and Control. Berkeley: University of California; 2019

[42] Liu D, Hong Y, Luo S, He X, Wu Z, Wang J. Feedback-driven quantum otto-like engines. Physical Review A. 2025;**111**(1):012203

[43] Bennett CH, Brassard G, Ekert AK. Quantum cryptography: Public key distribution and coin tossing. In: Proceedings of IEEE International Conference on Computers, Systems and Signal Processing. Bangalore, India; 1993. pp. 175-179

[44] Kosloff R, Levy A. Quantum heat engines and refrigerators: Continuous devices. Annual Review of Physical Chemistry. 2014;**65**:365

[45] Nakamura Y, Chief GL, Tsai JS. Coherent control of a single quantum bit. Nature. 1999;**398**(6726):786-788

[46] Nakamura Y, Tsai JS. Quantum coherence in a superconducting single-electron box. In: Exploring the Quantum/Classical Frontier: Recent Advances in Macroscopic Quantum Phenomena. 2003. pp. 97-129

[47] Van Der Wal CH, Ter Haar ACJ, Wilhelm FK, Schouten RN, Harmans CJPM, Orlando TP, et al. Quantum superposition of macroscopic persistent-current states. Science. 2000;**290**(5492): 773-777

Chapter 4

Quantum Speedup in Molecular Integral Evaluation through Angular Momentum Coupling

Hang Hu, Gilles Peslherbe, Hsu Kiang Ooi and Anguang Hu

Abstract

This chapter introduces a novel computational formalism utilizing Solid Harmonic Gaussian Orbitals (SHGOs) for efficiently calculating molecular integrals. Despite the prevalent utilization of Cartesian Gaussian Orbitals (CGOs) attributed to their straightforward nature, they inadequately capitalize on the intrinsic rotational symmetry characteristic of quantum systems. SHGOs, in contrast, clearly separate radial and angular components and align naturally with angular momentum symmetries. This alignment significantly enhances computational efficiency and accuracy in quantum chemical calculations. The approach described here leverages advanced angular momentum algebra techniques, particularly vector-coupling and vector-uncoupling schemes, to systematically reduce the complexity of molecular integral evaluations. By leveraging Clebsch-Gordan coefficients and Wigner 3-j symbols, the developed formalism significantly enhances computational efficiency over conventional CGO methods, as evidenced by a notable decrease in floating-point operations. Moreover, this SHGO-based method offers promising connections to emerging quantum computing applications. Quantum computing architectures inherently employ angular momentum entanglement principles, rendering SHGOs exceptionally well-suited for native quantum simulations. Specifically, angular momentum entanglement networks constructed from Clebsch-Gordan transformations capture rotational properties effectively, enabling scalable quantum simulation platforms and enhancing computational accuracy. Additionally, this methodology has significant implications beyond traditional quantum chemistry, offering pathways for quantum sensing and imaging technologies through the explicit consideration of orbital angular momentum. This integrated framework thus positions SHGO-based methods at the forefront of computational quantum chemistry, providing a critical tool to bridge classical methodologies with quantum computational advancements.

Keywords: solid harmonic Gaussian orbitals, quantum chemistry, angular momentum entanglement, Clebsch-Gordan coefficients, molecular integrals, quantum computing

1. Introduction

The field of computational quantum chemistry emerged from the pioneering efforts to solve Schrödinger's equation in the early twentieth century, aiming to accurately predict electronic structures and chemical properties [1–3]. The introduction of the Hartree-Fock method marked a significant advancement in the computation of electronic structures. This method systematically approximates electron-electron interactions through a mean-field approach [4–6]. Subsequently, density-functional theory (DFT) offered an alternative by concentrating on electron density instead of directly addressing wavefunctions, thereby facilitating more practical calculations for larger molecular systems [7–9]. In recent decades, Kohn-Sham density functional theory (KS-DFT) has become one of the most widely used quantum mechanical modeling methods, particularly for computing ground-state energies in periodic systems [7–9]. KS-DFT simplifies the solution of the Schrödinger equation by reducing its complexity from a 3 N problem, where N is the number of electrons to an electron density map. The key insight of KS-DFT is the recognition that many properties of materials can be accurately determined from electron density alone, thereby greatly simplifying computations.

In the time-independent Schrödinger equation [10–12], when the Hamiltonian operator $\widehat{H}$ acts on the stationary electronic wavefunction Ψ, the result is the stationary state energy E multiplied by the wavefunction Ψ, as expressed by:

$$\widehat{H}\Psi = E\Psi \tag{1}$$

Central to many computational methods in quantum chemistry is the representation of the wavefunction using a set of basis functions. Basis functions mathematically describe the spatial probability distribution of electrons around atoms and molecules [13–16]. Typically, these functions have two distinct components: a radial part, which describes electron distribution as a function of distance from atomic nuclei, and an angular part, which is an eigenfunction of the angular momentum operator. This structured representation forms the foundational framework upon which modern computational chemistry methods are constructed.

Systems exhibiting rotational symmetry particularly benefit from angular momentum considerations, as angular momentum algebra can greatly simplify the otherwise intricate mathematical formulations encountered in quantum chemistry [17–19]. Complex integrals can often be decomposed into two distinct types of terms. One set is rotationally invariant and directly dependent on the physical characteristics of the quantum system. The second set solely depends on the system's rotational symmetry and can be systematically expressed using angular momentum algebra [17–19]. These algebraic techniques, refined over decades, have proven invaluable across various domains, including atomic spectroscopy, molecular interactions, and nuclear reaction studies [4, 20].

In contemporary computational quantum chemistry, Cartesian Gaussian Orbitals (CGOs) have historically dominated computational implementations due to their mathematical simplicity and straightforward numerical integration [21–25]. CGOs simplify integral evaluations, making them particularly attractive for computational algorithms. Typically, calculations using CGOs are carried out extensively, with a conversion to Solid Harmonic Gaussian Orbitals (SHGOs) performed only in the final evaluation steps of integrals. Although computationally convenient, CGOs inherently misalign with the intrinsic rotational symmetry of quantum systems, leading to inefficiencies and redundancies when modeling spherical symmetry. In contrast, SHGOs present a sophisticated alternative explicitly engineered to leverage spherical symmetry. SHGOs clearly separate the radial and angular dependencies, with the angular component directly corresponding to eigenfunctions of the angular momentum operator. This intrinsic alignment with rotational symmetry significantly streamlines computational processes, particularly in systems where spherical symmetry is prominent. Despite their evident theoretical advantages, SHGOs have predominantly been investigated for one-electron molecular integrals, resulting in the relatively underdeveloped application of SHGOs in the domain of more complex two-electron molecular integrals [26–31].

Addressing this computational gap, this chapter introduces a novel SHGO-based formalism designed explicitly for the efficient calculation of molecular integrals. Our approach leverages advanced angular momentum algebra techniques—specifically vector-coupling and vector-uncoupling methods—to exploit symmetry properties systematically, resulting in markedly simplified and computationally efficient integral evaluations. The fundamental advantage arises from the direct coupling and uncoupling mechanisms inherent in the angular momentum structure of SHGOs, enabling substantial reductions in computational complexity. Furthermore, adopting SHGO methods provides promising pathways toward quantum computing applications. Quantum computing architectures inherently leverage angular momentum coupling principles, making SHGOs particularly suitable for quantum simulations. By integrating SHGO methodologies within quantum computational frameworks, significant advancements in computational accuracy and efficiency can be realized, positioning SHGOs as a valuable bridge between classical computational chemistry and emerging quantum technologies.

2. Molecular integrals

In general, SHGOs are defined as

$$\chi_{m_a}^{l_a \vec{a} \alpha}\left(\vec{r}\right) = N(l_a, \alpha) Y_{m_a}^{l_a}\left(\vec{r} - \vec{a}\right) e^{-\alpha\left|\vec{r} - \vec{a}\right|^2} \tag{2}$$

where $\vec{a}$ is the orbital atomic center, α the Gaussian exponent, $N(l_a, \alpha)$the normalization constant, and $Y_{m_a}^{l_a}\left(\vec{r} - \vec{a}\right)$ are solid harmonics, solutions of the Laplace equation. Solid harmonics are related to the spherical harmonics as [18].

$$Y_m^l\left(\vec{r}\right) = \sqrt{\frac{4\pi}{2l+1}} r^l Y_{lm}(\theta, \phi) \tag{3}$$

where $Y_{lm}(\theta, \phi)$ are the spherical harmonics with the phase convention of Condon and Shortley [4], θ is the polar angle, ϕ is the azimuthal angle, l and m are the orbital angular momentum and magnetic quantum numbers, respectively, r is the radial distance from the origin. By applying the Hobson theorem of solid harmonics [32] SHGOs can be further simplified through the use of solid harmonic derivatives with respect to the orbital atomic center,

$$\chi_{m_a}^{l_a \vec{a} \alpha}\left(\vec{r}\right) = N(l_a, \alpha) \left(\frac{1}{2\alpha}\right)^l Y_{m_a}^{l_a}(\nabla_{\vec{a}}) e^{-\alpha\left|\vec{r} - \vec{a}\right|^2} \tag{4}$$

Without the normalization constants, the general expression for a molecular overlap integral is of the form

$$S_{\alpha\beta}^{\vec{a}\vec{b}}(l_a m_a, l_b m_b) = \left(\frac{1}{2\alpha}\right)^{l_a}\left(\frac{1}{2\beta}\right)^{l_b}\int Y_{m_a}^{l_a}\left(\vec{r}-\vec{a}\right)Y_{m_b}^{l_b}\left(\vec{r}-\vec{b}\right)e^{-\alpha\left|\vec{r}-\vec{a}\right|^2}e^{-\beta\left|\vec{r}-\vec{b}\right|^2}d\tau \quad (5)$$

$S_{\alpha\beta}^{\vec{a}\vec{b}}(l_a m_a, l_b m_b)$ is the overlap between two Gaussian-type orbitals centered at $\vec{a}$ and $\vec{b}$, each characterized by angular quantum numbers $l_a m_a, l_b m_b$, respectively. α and β are the Gaussian exponential parameters. Using the Gaussian product rule, we can shift the center from $\vec{a}$ and $\vec{b}$ and bycenter at $\vec{P}$

$$e^{-\alpha\left|\vec{r}-\vec{a}\right|^2}e^{-\beta\left|\vec{r}-\vec{b}\right|^2} = e^{-\frac{\alpha\beta}{\alpha+\beta}\left|\vec{a}-\vec{b}\right|^2}e^{-(\alpha+\beta)\left|\vec{r}-\vec{P}\right|^2} \quad (6)$$

where $\vec{P} = \frac{\alpha\vec{a}+\beta\vec{b}}{\alpha+\beta}$, and integration yields overlap integrals given by

$$S_{\alpha\beta}^{\vec{a}\vec{b}}(l_a m_a, l_b m_b) = \left(\frac{1}{2\alpha}\right)^{l_a}\left(\frac{1}{2\beta}\right)^{l_b}\left(\frac{\pi}{\alpha+\beta}\right)^{\frac{3}{2}}\int Y_{m_a}^{l_a}\left(\vec{r}-\vec{a}\right)Y_{m_b}^{l_b}\left(\vec{r}-\vec{b}\right)e^{-\frac{\alpha\beta}{\alpha+\beta}\left|\vec{a}-\vec{b}\right|^2}d\tau \quad (7)$$

The overlap integral in Eq. (7) can be evaluated by applying the differential and product rules of solid harmonic derivatives [27]. This approach is similar to the previously derived vector-uncoupling scheme of angular momentum because there are no terms related to the $l_a + l_b$quantum number [27]. However, this is suboptimal in terms of computational efficiency, with too many harmonic derivatives acting on different atomic centers, resulting in deep angular momentum coupling of the atomic orbitals. Our new approach eliminates the harmonics derivatives with the addition of angular momentum [18]. Following the addition theorem, we can shift orbital atomic centers of the solid harmonics from $\vec{a}$ and $\vec{b}$ to the same center $\vec{P}$ of the Gaussian $e^{-(\alpha+\beta)\left|\vec{r}-\vec{P}\right|^2}$, and obtain the following equation

$$Y_{m_a}^{l_a}\left(\vec{r}-\vec{a}\right) = \sum_{l_1=0}^{l_a}\left(\frac{\beta}{\alpha+\beta}\right)^{l_a-l_1}\sum_{m_1=-l_1}^{l_1}\varepsilon_{l_a l_1 m_a m_1}Y_{m_a-m_1}^{l_a-l_1}\left(\vec{a}-\vec{b}\right)Y_{m_1}^{l_1}\left(\vec{r}-\vec{P}\right) \quad (8)$$

Here, the vector-coupling coefficients of angular momentum are given by [33].

$$\varepsilon_{l_1 l_2 m_1 m_2} = \sqrt{\binom{l_1+m_1}{l_2+m_2}\binom{l_1-m_1}{l_2-m_2}} \quad (9)$$

and are also Clebsch-Gordan (CG) transformation coefficients [26]. Therefore, a two-center overlap integral transforms into a single-center integral,

$$S_{\alpha\beta}^{\vec{a}\vec{b}}(l_a m_a, l_b m_b) = e^{-\frac{\alpha\beta}{\alpha+\beta}\left|\vec{a}-\vec{b}\right|^2}\sum_{l_1=0}^{l_a}\left(\frac{\beta}{\alpha+\beta}\right)^{l_a-l_1}$$

$$\sum_{m_1=-l_1}^{l_1}\varepsilon_{l_a l_1 m_a m_1}Y_{m_a-m_1}^{l_a-l_1}\left(\vec{a}-\vec{b}\right)\sum_{l_2=0}^{l_b}\left(\frac{\alpha}{\alpha+\beta}\right)^{l_b-l_2}$$

$$\sum_{m_2=-l_2}^{l_2}\varepsilon_{l_b l_2 m_b m_2}Y_{m_b-m_2}^{l_b-l_2}\left(\vec{b}-\vec{a}\right)\int Y_{m_1}^{l_1}\left(\vec{r}-\vec{P}\right)Y_{m_2}^{l_2}\left(\vec{r}-\vec{P}\right)e^{-(\alpha+\beta)\left|\vec{r}-\vec{P}\right|^2}d\tau \quad (10)$$

Using Eq. (3), we can convert the solid harmonics in the last line of Eq. (10) into spherical harmonics.

$$\int Y_{m_1}^{l_1}\left(\vec{r}-\vec{p}\right)Y_{m_2}^{l_2}\left(\vec{r}-\vec{p}\right)e^{-(\alpha+\beta)\left|\vec{r}-\vec{p}\right|^2}d\tau =$$

$$\sqrt{\frac{4\pi}{2l_1+1}}\sqrt{\frac{4\pi}{2l_2+1}}\int\left(\vec{r}-\vec{p}\right)^{l_1}\left(\vec{r}-\vec{p}\right)^{l_2}Y_{m_1}^{l_1}(\theta,\varphi)Y_{m_2}^{l_2}(\theta,\varphi)e^{-(\alpha+\beta)\left|\vec{r}-\vec{p}\right|^2}d\tau \quad (11)$$

Spherical harmonics at the same centers are orthogonal and the integrals of spherical harmonics can be represented by Kronecker delta symbols,

$$\int Y_{m_1}^{l_1}\left(\vec{r}-\vec{p}\right)Y_{m_2}^{l_2}\left(\vec{r}-\vec{p}\right)e^{-(\alpha+\beta)\left|\vec{r}-\vec{p}\right|^2}d\tau =$$

$$\sqrt{\frac{4\pi}{2l_1+1}}\sqrt{\frac{4\pi}{2l_2+1}}\delta_{l_1l_2}\delta_{m_1m_2}\int_0^{\infty}\left|\vec{r}-\vec{p}\right|^{2l_1+2}e^{-(\alpha+\beta)\left|\vec{r}-\vec{p}\right|^2}d\left(\vec{r}-\vec{p}\right) \quad (12)$$

In the end the two center integral of Eq. (10) can be expressed as a collection of constants,

$$\int Y_{m_1}^{l_1}\left(\vec{r}-\vec{p}\right)Y_{m_2}^{l_2}\left(\vec{r}-\vec{p}\right)e^{-(\alpha+\beta)\left|\vec{r}-\vec{p}\right|^2}d\tau = \sqrt{\frac{4\pi}{2l_1+1}}\sqrt{\frac{4\pi}{2l_2+1}}\delta_{l_1l_2}\delta_{m_1m_2}\frac{(2l_1+1)!}{[2(\alpha+\beta)]^{l_1+1}}\sqrt{\frac{\pi}{2}} \quad (13)$$

By inserting Eq. (13) back into Eq. (10), we have:

$$S_{\alpha\beta}^{\vec{a}\vec{b}}(l_am_a;l_bm_b) = 4\pi\left(\frac{\pi}{2}\right)^{\frac{1}{2}}e^{-\frac{\alpha\beta}{\alpha+\beta}\left|\vec{a}-\vec{b}\right|^2}\sum_{l_1=0}^{\min(l_a,l_b)}\left(\frac{\beta}{\alpha+\beta}\right)^{l_a-l_1}\left(\frac{\alpha}{\alpha+\beta}\right)^{l_b-l_1}\left(\frac{1}{2l_1+1}\right)(-1)^{l_b-l_1}$$

$$\frac{(2l_1+1)!!}{[2(\alpha+\beta)]^{\frac{2l_1+3}{2}}}\sum_{m_1=-l_1}^{l_1}\varepsilon_{l_al_1m_am_1}\varepsilon_{l_bl_1m_bm_1}Y_{m_a-m_1}^{l_a-l_1}\left(\vec{a}-\vec{b}\right)Y_{m_b-m_1}^{l_b-l_1}\left(\vec{a}-\vec{b}\right) \quad (14)$$

The summation in the second line of Eq. (14) is independent of orbitals and determined only by the rotational properties of angular momentum for an overlap integral. The last line of Eq. (14) is an elegant and remarkably efficient solution to demonstrate that quantum angular momenta of the same atomic center cannot entangle with each other due to their orthonormality. This indicates that if a quantum system does not feature much coupling, it will be much easier to simulate.

In the same way, the nuclear Coulomb attraction integrals can be directly calculated by simply introducing a nuclear Coulomb operator located at the center $\vec{c}$, into the integral of the last line of Eq. (14),

$$N_{\alpha\beta}^{\vec{a}\vec{b}\vec{c}}(l_am_a,l_bm_b) = e^{-\frac{\alpha\beta}{\alpha+\beta}\left|\vec{a}-\vec{b}\right|^2}\sum_{l_1=0}^{l_a}\left(\frac{\beta}{\alpha+\beta}\right)^{l_a-l_1}\sum_{m_1=-l_1}^{l_1}\varepsilon_{l_al_1m_am_1}Y_{m_a-m_1}^{l_a-l_1}\left(\vec{a}-\vec{b}\right)$$

$$\sum_{l_2=0}^{l_b}\left(\frac{\alpha}{\alpha+\beta}\right)^{l_b-l_2}\sum_{m_2=-l_2}^{l_2}\varepsilon_{l_bl_2m_bm_2}Y_{m_b-m_2}^{l_b-l_2}\left(\vec{b}-\vec{a}\right)\int\frac{Y_{m_1}^{l_1}\left(\vec{r}-\vec{P}\right)Y_{m_2}^{l_2}\left(\vec{r}-\vec{P}\right)e^{-(\alpha+\beta)\left|\vec{r}-\vec{P}\right|^2}}{\left|\vec{r}-\vec{c}\right|}d\tau \quad (15)$$

The integral on the last line of Eq. (15) clearly expresses the nuclear Coulomb attraction integral as the interaction between Gaussian-distributed angular momenta at the center $\vec{P}$ and the nuclear Coulomb potential at the center $\vec{c}$. The well-known formula of solid harmonics products may be applied to simplify the calculation,

$$Y_{m_1}^{l_1}\left(\vec{r}-\vec{P}\right)Y_{m_2}^{l_2}\left(\vec{r}-\vec{P}\right)=\sum_{|l_1-l_2|}^{l_1+l_2}\sum_{m_p=-l_p}^{l_p}\left[\frac{(2l_1+1)(2l_1+1)}{4\pi(2l_p+1)}\right]$$
$$\begin{pmatrix} l_1\ l_1\ l_p \\ 0\ 0\ 0 \end{pmatrix}\begin{pmatrix} l_1\ l_2\ l_p \\ m_1 m_2 m_p \end{pmatrix}\left|\vec{r}-\vec{P}\right|^{l_1+l_2-l_p}Y_{m_p}^{l_p}\left(\vec{r}-\vec{P}\right) \tag{16}$$

where $\begin{pmatrix} l_1\ l_2\ l_p \\ m_1 m_2 m_p \end{pmatrix}$ is the Wigner 3-j symbol. Note that the only $l_p$values on the right should satisfy $l_1+l_2+l_p=$ evenintegers. Combining CG coefficients $\varepsilon_{l_a l_1 m_a m_1}, \varepsilon_{l_b l_2 m_b m_2}, Y_{m_a-m_1}^{l_a-l_1}\left(\vec{a}-\vec{b}\right)$ and $Y_{m_b-m_2}^{l_b-l_2}\left(\vec{b}-\vec{a}\right)$ in Eq. (15) with the Wigner 3-j symbols, we have a three-dimensional array $(l_1+l_2,l_p,m_p)_{\vec{b}l_b m_b}^{\vec{a}l_a m_a}$, which is independent of orbitals. Thus, the nuclear Coulomb integral can be further simplified to

$$\int\frac{|\mathrm{r}-\mathrm{P}|^{2\left[\frac{l_1+l_2-l_p}{2}\right]}Y_{m_p}^{l_p}\left(\vec{r}-\vec{P}\right)e^{-(\alpha+\beta)\left|\vec{r}-\vec{P}\right|^2}}{\left|\vec{r}-\vec{c}\right|}d\tau$$
$$=\left(-\frac{d}{d(\alpha+\beta)}\right)^{\left[\frac{l_1+l_2-l_p}{2}\right]}\int\frac{Y_{m_p}^{l_p}\left(\vec{r}-\vec{P}\right)e^{-(\alpha+\beta)\left|\vec{r}-\vec{P}\right|^2}}{\left|\vec{r}-\vec{c}\right|}d\tau$$
$$=2\pi Y_{m_p}^{l_p}\left(\vec{c}-\vec{P}\right)\left(-\frac{d}{d(\alpha+\beta)}\right)^{\left[\frac{l_1+l_2-l_p}{2}\right]}\int_0^1 du u^{2l_p}\frac{e^{-(\alpha+\beta)u^2\left|\vec{c}-\vec{P}\right|^2}}{\alpha+\beta} \tag{17}$$

where $\left[\frac{l_1+l_2-l_p}{2}\right]$ is a positive integer. Compared to all available integrals of the same type with either SHGOs or CGOs, Eq. (16) is the simplest expression for general nuclear Coulomb attraction integrals. The final expression of Eqs. (18) combines Eqs. (14), (15) and (16):

$$N_{\alpha\beta}^{\vec{a}\vec{b}\vec{c}}(l_a m_a, l_b m_b)=e^{-\frac{\alpha\beta}{\alpha+\beta}\left|\vec{a}-\vec{b}\right|^2}\sum_{l_1=0}^{l_a}\left(\frac{\beta}{\alpha+\beta}\right)^{l_a-l_1}\sum_{m_1=-l_1}^{l_1}\varepsilon_{l_a l_1 m_a m_1}Y_{m_a-m_1}^{l_a-l_1}\left(\vec{a}-\vec{b}\right)$$
$$\sum_{l_2=0}^{l_b}\left(\frac{\alpha}{\alpha+\beta}\right)^{l_b-l_2}\sum_{m_2=-l_2}^{l_2}\varepsilon_{l_b l_2 m_b m_2}Y_{m_b-m_2}^{l_b-l_2}\left(\vec{b}-\vec{a}\right)$$
$$\sum_{|l_1-l_2|}^{l_1+l_2}\sum_{m_p=-l_p}^{l_p}\left[\frac{(2l_1+1)(2l_1+1)}{4\pi(2l_p+1)}\right]\begin{pmatrix} l_1\ l_1\ l_p \\ 0\ 0\ 0 \end{pmatrix}\begin{pmatrix} l_1\ l_2\ l_p \\ m_1 m_2 m_p \end{pmatrix}2\pi Y_{m_p}^{l_p}\left(\vec{c}-\vec{P}\right)$$
$$\left(-\frac{d}{d(\alpha+\beta)}\right)^{\left[\frac{l_1+l_2-l_p}{2}\right]}\int_0^1 du u^{2l_p}\frac{e^{-(\alpha+\beta)u^2\left|\vec{c}-\vec{P}\right|^2}}{\alpha+\beta} \tag{18}$$

For a general two-electron Coulomb integral, there are four Gaussian exponents α, β, δ, and λ located at four atomic orbital centers $\vec{a}$, $\vec{b}$, $\vec{c}$, and $\vec{d}$ respectively. We can transform this two-electron, four-center Coulomb integral into a two-center one with the centres located at $\vec{P} = \frac{\alpha\vec{a}+\beta\vec{b}}{\alpha+\beta}$ and $\vec{Q} = \frac{\delta\vec{c}+\lambda\vec{d}}{\delta+\lambda}$ in terms of four Gaussian-distributed angular momenta based on Eq. (6). This leads to the following simple expression

$$I\left(\vec{r}_1 - \vec{P}, \vec{r}_2 - \vec{Q}\right)^{\xi l_1 m_1 l_2 m_2}_{\varsigma l_3 m_3 l_4 m_4} = \iint \frac{Y^{l_1}_{m_1}\left(\vec{r}_1 - \vec{P}\right) Y^{l_2}_{m_2}\left(\vec{r}_1 - \vec{P}\right) Y^{l_3}_{m_3}\left(\vec{r}_2 - \vec{Q}\right) Y^{l_4}_{m_4}\left(\vec{r}_2 - \vec{Q}\right) e^{-\xi\left|\vec{r}_1 - \vec{P}\right|^2} e^{-\varsigma\left|\vec{r}_2 - \vec{Q}\right|^2}}{\left|\vec{r}_1 - \vec{r}_2\right|} d\tau_1 d\tau_2 \tag{19}$$

We use the addition theorem shown previous in Eq. (8), to combine the product of two solid harmonics into a summation of a single solid harmonic.

$$Y^{l_3}_{m_3}\left(\vec{r}_2 - \vec{Q}\right) Y^{l_4}_{m_4}\left(\vec{r}_2 - \vec{Q}\right) = \sum_{l=|l_3-l_4|}^{l_3+l_4} \sum_{m_p=-l}^{l} \left[\frac{(2l_3+1)(2l_4+1)(2l+1)}{4\pi}\right] \begin{pmatrix} l_3 & l_4 & l \\ 0 & 0 & 0 \end{pmatrix} \begin{pmatrix} l_3 & l_4 & 1 \\ m_3 & m_4 & m \end{pmatrix} \left|\vec{r}_2 - \vec{Q}\right|^{l_3+l_4-l} Y^{l}_{m}\left(\vec{r}_2 - \vec{Q}\right)$$

Applying the simplification to Eq. (19), we get

$$I\left(\vec{r}_1 - \vec{P}, \vec{r}_2 - \vec{Q}\right)^{\xi l_p m_p}_{\varsigma l_q m_q} = \left(-\frac{d}{d\xi}\right)^{2\left[\frac{l_1+l_2-l_p}{2}\right]} \left(-\frac{d}{d\varsigma}\right)^{2\left[\frac{l_3+l_4-l_q}{2}\right]} \iint \frac{Y^{l_p}_{m_p}\left(\vec{r}_1 - \vec{P}\right) Y^{l_q}_{m_q}\left(\vec{r}_2 - \vec{Q}\right) e^{-\xi\left|\vec{r}_1 - \vec{P}\right|^2} e^{-\varsigma\left|\vec{r}_2 - \vec{Q}\right|^2}}{\left|\vec{r}_1 - \vec{r}_2\right|} d\tau_1 d\tau_2 \tag{20}$$

Lastly, the denominator of the equation can be expressed as a product of spherical harmonics.

$$\frac{1}{\left|\vec{r}_1 - \vec{r}_2\right|} = \frac{1}{\left|\left(\vec{r}_1 - \vec{Q}\right) - \left(\vec{r}_2 - \vec{Q}\right)\right|} = \sum_{l=0}^{\infty} \sum_{m=-l}^{l} \frac{4\pi}{2l+1} \left(\frac{\left|\vec{r} - \vec{Q}\right|^{l}_{<}}{\left|\vec{r} - \vec{Q}\right|^{l+1}_{>}}\right) Y^{*}_{lm}(\theta_2, \varphi_2) Y_{lm}(\theta_1, \varphi_1) \tag{21}$$

In this simplified process, there are two three-dimensional arrays $(l_1 + l_2, l_p, m_p)^{\vec{a} l_a m_a}_{\vec{b} l_b m_b}$ and $(l_3 + l_4, l_q, m_q)^{\vec{c} l_c m_c}_{\vec{d} l_d m_d}$, which are independent of Gaussian exponents and are common to all relevant integrals including nuclear Coulomb ones. As a result, $I\left(\vec{r}_1 - \vec{P}, \vec{r}_2 - \vec{Q}\right)^{\xi l_p m_p}_{\varsigma l_q m_q}$ reduces to

$$I\left(\vec{r}_1-\vec{P},\vec{r}_2-\vec{Q}\right)^{\xi l_p m_p}_{\varsigma l_q m_q}=2\pi^{\frac{5}{2}}\left(-\frac{d}{d\xi}\right)^{2\left[\frac{l_1+l_2-l_p}{2}\right]}\left(-\frac{d}{d\varsigma}\right)^{2\left[\frac{l_3+l_4-l_q}{2}\right]}\frac{1}{\xi\varsigma\sqrt{\xi+\varsigma}}\left(\frac{\varsigma}{\xi+\varsigma}\right)^{l_p}$$
$$\left(\frac{\xi}{\xi+\varsigma}\right)^{l_q}\sum\nolimits_{l_t=0}^{min\left(l_p,l_q\right)}\left(-\frac{\xi+\varsigma}{2\xi\varsigma}\right)^{l_t}(2l_t-1)!!\int_0^1 u^{2\left(l_p+l_q-l_t\right)}e^{-\frac{\xi\varsigma}{\xi+\varsigma}\left|\vec{P}-\vec{Q}\right|^2u^2}du \tag{22}$$
$$\sum\nolimits_{m_t=-l_t}^{l_t}\varepsilon_{l_p l_t m_p m_t}\varepsilon_{l_q l_t m_q m_t}Y^{l_q-l_t}_{m_q-m_t}\left(\vec{P}-\vec{Q}\right)Y^{l_p-l_t}_{m_p-m_t}\left(\vec{Q}-\vec{P}\right)$$

3. Computational implementation

The implementation of nuclear attraction integral in Eq. (15) can be split into radial and angular component. The first one is a three-dimensional array $(l_1+l_2,l_p,m_p)^{\vec{a}l_a m_a}_{\vec{b}l_b m_b}$ and is independent of Gaussian exponents, which makes it possible to calculate it once and reuse it in later steps, and even applicable to two-electron Coulomb integrals. The second part involves $Y^{l_p}_{m_p}\left(\vec{c}-\vec{P}\right)$, which is dependent on Gaussian orbitals. The third part is related to the Boys function, $\left(-\frac{d}{d(\alpha+\beta)}\right)^{\left[\frac{l_1+l_2-l_p}{2}\right]}\int_0^1 duu^{2l_p}\frac{e^{-(\alpha+\beta)u^2\left|\vec{c}-\vec{P}\right|^2}}{\alpha+\beta}$ dependent on Gaussian exponents [34], which can be simplified as $\left(\frac{d}{dx}\right)^n\frac{e^{-x}}{x}$ and solved using the recursion formulae derived from the modified spherical Bessel function of the second kind [35].

```
def BoysFunction(nu, x):
      if x < 1e-7:
         return (2*nu+1)**(-1) - x*(2*nu+3)**(-1)
   return 0.5 * x**(-(nu+0.5)) * gamma(nu+0.5) * gammainc(nu+0.5, x)
def bessel_recursion(n, z):
   # Base cases
   if n == 0:
      return 1 / z # f_0(z) = z^(-1)
   elif n == 1:
      return (z + 1) / (z**2) # f_1(z) = (z+1)/z^2
   # Recursive formula: f_n(z) = f_(n-2)(z) + (2n-1) * z^(-1) * f_(n-1)(z)
   return bessel_recursion(n-2, z) + (2*n-1) * (1/z) * bessel_recursion(n-1, z)
```

The main code is a six-level nested loop structure that computes all allowed angular momentum couplings then projected onto the spherical harmonics. An example of the loops are given below.

```
def nuclear_shgo(...):
      for l1 in range(la + 1):
         for m1 in range(-l1, l1 + 1):
            for l2 in range(lb + 1):
               for m2 in range(-l2, l2 + 1):
                  for lp in range(abs(l1 - l2), l1 + l2 + 1):
                     for mp in range(-lp, lp + 1): ...
```

This implementation captures the essence of SHGO-based methods: efficient exploitation of angular momentum symmetry. By explicitly decomposing angular contributions through Clebsch-Gordan coefficients and Wigner 3-j symbols, the approach ensures only symmetry-allowed terms are evaluated. The Boys function, combined with the recursive Bessel approximation, effectively handles the radial decay and accurately addresses the nuclear singularity. The most computationally intensive steps involve evaluating Wigner symbols and the solid harmonic terms across all valid angular momenta. However, these costs are mitigated by careful adherence to triangle conditions, which skip disallowed angular combinations. **Figure 1** shows a comparison of the computational cost in terms of floating point operations for evaluating the nuclear Coulomb attraction integral with general CGOs scheme and SHGOs [36].

The integration of a single primitive Cartesian Gaussian function necessitates nine loops, which constitute the primary computational expense of the CGO scheme shown of **Figure 1**. Previous SHGO methods exploiting only the vector-uncoupling scheme of angular momentum, as implemented in the ParaGauss package, have a similar computational cost [31]. This relatively high cost is the result of three-fold summations over angular momentum l and magnetic m quantum numbers [28]. The interaction of two orbital angular momenta from different centers with a nuclear Coulomb potential can lead to significant quantum coupling under the vector-uncoupling scheme. As a result, the vector-uncoupling scheme of angular momentum is not suitable for multi-center molecular integrals. This may partially elucidate why molecular integrals, particularly those involving SHGOs, were not integrated into earlier quantum chemistry software packages. When compared against the standard nuclear attraction integral for CGOs, Eq. (15) has more Wigner 3-j and CG coefficient calculation steps but only 6 loops, and this is where the bulk of the computational saving comes from. The number of floating-point operations for the evaluation of Eq. (15) scheme is about 3.6 times less than the corresponding CGO nuclear attraction integrals (**Table 1**).

The two-electron Coulomb integral is shown in Eq. (22), it involves two 5-dimensional array factors. The first array $(l_1+l_2, l_p, l_3+l_4, l_q, l_t)_{\vec{Q}\varsigma}^{\vec{P}\xi}$ corresponds to the first two lines of Eq. (18). The second array $(l_p, m_p, l_q, m_q, l_t)_{\vec{Q}}^{\vec{P}}$ is due to the summation over m_t in the last line of Eq. (18). The final expression of the general two-electron Coulomb integral is given by the summation of four array factors, $(l_1+l_2, l_p, m_p)_{\vec{b}l_bm_b}^{\vec{a}l_am_a}$, $(l_3+l_4, l_q, m_q)_{\vec{d}l_dm_d}^{\vec{c}l_cm_c}$,

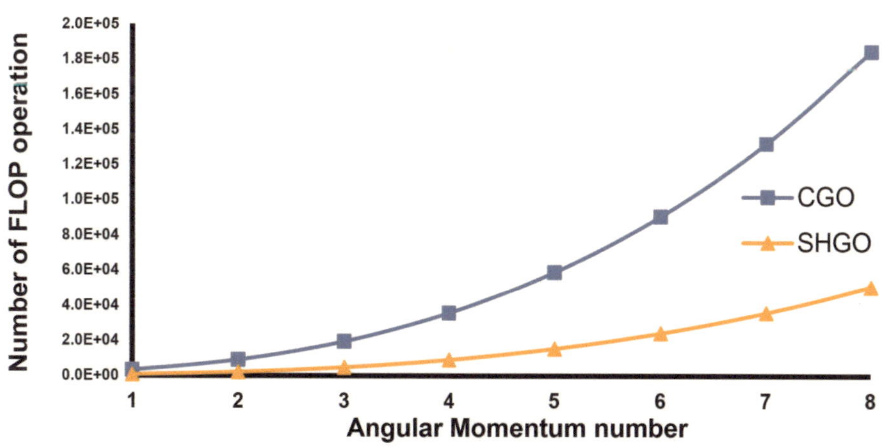

Figure 1.
Computational cost in terms of floating point operations in the calculation of nuclear attraction Coulomb integrals as a function of the highest orbital angular momentum number with CGOs versus SHGOs [36].

Calculation steps with CGOs [37]	Cost	Calculation steps with SHGOs [28]	Cost	This work with SHGOs	Cost
Boys functions	LP^2	Solid harmonics	L^2	Solid harmonics	L^2P^2
Hermite integrals	L^4P^2	Solid harmonics	L^2P^2	Boys functions	LP^2
Expansion coefficients	L^2P^2	Differential rule	L^4P^2		
Cartesian integrals	L^7P^2	Product rule	L^6P^2		
Primitive contraction	L^3P^2	Boys functions	LP^2		
Solid harmonics	L^5				

L is the highest orbital angular momentum number of the Gaussian orbitals and P is the number of primitive Gaussian functions.

Table 1.
The cost associated with computing nuclear attraction integrals.

$(l_p, m_p, l_q, m_q, l_t)_{\vec{Q}}^{\vec{P}}$, and $(l_1 + l_2, l_p, l_3 + l_4, l_q, l_t)_{\varsigma\vec{Q}}^{\xi\vec{P}}$, with a scaling of $L^7P^4$where L is the highest angular momentum quantum number and P is the number of primitive Gaussian functions. The actual scaling may be reduced when we apply the constraints where $l_1 + l_2 + l_p$ and $l_3 + l_4 + l_q$ must be even positive integers. The constraints also apply to the summation in the second line of Eq. (18), together with$|l_1 - l_2| \le l_p \le l_1 + l_2$ and $|l_3 - l_4| \le l_q \le l_3 + l_4$. Moreover, symmetry conditions for the Wigner 3-j symbol also reduce the number of nonzero terms in the array factors $(l_1 + l_2, l_p, m_p)_{\vec{b}l_bm_b}^{\vec{a}l_am_a}$ and $(l_3 + l_4, l_q, m_q)_{\vec{d}l_dm_d}^{\vec{c}l_cm_c}$, resulting in significant computational savings. In terms of implementation most of the utility functions, such as Boy's function, Solid harmonics and other angular prefactors are reusable from the nuclear integral. For our purposes, we use the Sympy library to perform the symbolic computation of radial derivatives with respect to the Gaussian exponents ξ and ς. These derivatives arise from contraction with higher-order radial basis functions. The symbolic method avoids approximation errors that may arise from finite-difference derivatives, while maintaining flexibility across a wide range of angular momentum and contraction configurations. The required number of derivatives with respect to each exponent is determined according to the specific angular and radial quantum numbers of the shell pair.

4. Quantum computing and future applications

Quantum computing represents a transformative potential for molecular modeling and simulation, addressing classical computational limitations such as the diagonalization of large two-body operators and the evaluation of complex two-body matrix elements encountered in quantum chemistry [38, 39]. Quantum algorithms, notably the Variational Quantum Eigensolver (VQE) and Quantum Phase Estimation (QPE), offer promising strategies for accurately evaluating molecular energies and properties [39–48]. Both methods involve key procedural steps: constructing the fermionic Hamiltonian, mapping this Hamiltonian onto qubit systems, and optimization procedures. Although VQE is more widely utilized, QPE has shown considerable promise due to its inherent precision.

$$\mathrm{H} = \mathrm{T} + \mathrm{V}$$

$$V \approx G = \frac{1}{2}\sum\nolimits_{p,q,r,s} G_{pqrs} a_p^{+} a_q a_r^{+} a_s \tag{23}$$

A critical challenge in quantum chemical computations lies in the diagonalization and evaluation of two-body Coulomb operators, significantly impacting computational scaling and complexity. Classical second quantization formalism results in large two-body operator matrices, complicating their efficient diagonalization. Recent advancements, such as tensor hypercontraction (THC) [49], approximate these complex four-center tensors as products of lower-dimensional tensors G, shown in Eq. (23). This approach simplifies computations significantly but introduces the challenge of determining hypercontraction parameters, typically derived via density fitting. However, density fitting itself remains a computational bottleneck, compounded by the non-orthogonal and non-unitary nature of traditional basis sets.

$$\chi_p^{l_{ia} m_{ia}}(\varphi_a, \theta_a) = c_p^{l_{ia} m_{ia}} Y_{l_{ia} m_{ia}}(\varphi_a, \theta_a) \tag{24}$$

Eq. (24) is an atom-centered angular momentum basis, which is constructed with a Fermion projection operator $c_p^{l_{ia} m_{ia}}$, and a spherical harmonic $Y_{l_{ia} m_{ia}}(\varphi_a, \theta_a)$ with angular momentum l_{ia} and m_{ia} centered at $\vec{a}$.

$$\mathrm{c}^{\dagger}_{l_{ia} m_{ia} \sigma}(\theta_a, \varphi_a) = \sum\nolimits_{p=1}^{\frac{N}{2}} \chi_p^{l_{ia} m_{ia}}(\theta_a, \varphi_a) a^{\dagger}_{p\sigma}$$

$$c_{l_{ia} m_{ia} \sigma}(\theta_a, \varphi_a) = \sum\nolimits_{p=1}^{\frac{N}{2}} \chi_p^{l_{ia} m_{ia}}(\theta_a, \varphi_a) a_{p\sigma} \tag{25}$$

This projection operator $c^{\dagger}_{l_{ia} m_{ia} \sigma}$ and $c_{l_{ia} m_{ia} \sigma}$ projects the original fermion ladder operators $a^{\dagger}_{p\sigma}$ $a_{p\sigma}$to the unitary angular momentum basis $\chi_p^{l_{ia} m_{ia}}(\theta_a, \varphi_a)$. This new approach to basis explicitly incorporates rotational symmetry through spherical harmonics. Projecting molecular orbitals onto atom-centered spherical harmonics ensures orthogonality and unitarity, substantially reducing complexity during diagonalization. The angular momentum basis can be obtained from decomposing the electron pair density $\phi_p(\vec{r})\phi_q(\vec{r})$ on a real grid,

$$\phi_p(\vec{r})\phi_q(\vec{r}) = \sum\nolimits_{a=1}^{N_a} \sum\nolimits_{ia=1}^{N_{ia}} \chi_p^{l_{ia} m_{ia}}(\varphi_a, \theta_a) \zeta_{m_{ia}}^{l_{ia}}(\vec{r} - \vec{a}) \chi_q^{l_{ia} m_{ia}}(\varphi_a, \theta_a) \tag{26}$$

where N_{ia} is the total number of basis functions centered at the atom $\vec{a}$ and $\zeta_{m_{ia}}^{l_{ia}}(\vec{r} - \vec{a})$ is a coulomb integrand associated with partitioning the electronic pair density centered at $\vec{a}$.

$$\zeta_{m_{ia}}^{l_{ia}}(\vec{r} - \vec{a}) = \frac{\rho_a(\vec{r} - \vec{a})}{\sum_{ja=1}^{N_{ia}} c_p^{l_{ja} m_{ja}} Y_{l_{ia} m_{ia}}(\theta_a, \varphi_a) c_q^{l_{ja} m_{ja}} Y_{l_{ia} m_{ia}}(\theta_a, \varphi_a)} \tag{27}$$

In this framework, the Coulomb operator is a summation of $n_{l_{ia} m_{ia}, \alpha} n_{l_{ib} m_{ib}, \beta}$, which are number operators acting on the quantum angular momentum basis.

$$G = \frac{1}{2}\sum_{\alpha,\beta\in\{\uparrow,\downarrow\}}\sum_{p,q,r,s=1}^{\frac{N}{2}} G_{pqrs} a^{\dagger}_{p,\alpha} a_{q,\alpha} a^{\dagger}_{r,\beta} a_{s,\beta}$$
$$= \frac{1}{2}\sum_{\alpha,\beta\in\{\uparrow,\downarrow\}}\sum_{a,b=1}^{N_a}\sum_{ia=1}^{N_{ia}}\sum_{ib=1}^{N_{ib}} \zeta^{l_{ia}m_{ia}}_{l_{ib}m_{ib}} n_{l_{ia}m_{ia},\alpha} n_{l_{ib}m_{ib},\beta} \tag{28}$$

The number operator is related to the fermion projection operations by:

$$n_{l_{ia}m_{ia},\alpha} = c^{\dagger}_{l_{ia}m_{ia}\sigma}(\theta_a, \varphi_a) c_{l_{ia}m_{ia}\sigma}(\theta_a, \varphi_a) \tag{29}$$

The final Coulomb operator is shown here:

$$G_{pqrs} = \sum_{a,b=1}^{N_a}\sum_{ia=1}^{N_{ia}}\sum_{ib=1}^{N_{ib}} \chi_p^{l_{ia}m_{ia}}(\varphi_a, \theta_a)\chi_q^{l_{ia}m_{ia}}(\varphi_a, \theta_a)\zeta^{l_{ia}m_{ia}}_{l_{ib}m_{ib}}\chi_r^{l_{ib}m_{ib}}(\varphi_b, \theta_b)\chi_s^{l_{ib}m_{ib}}(\varphi_b, \theta_b)$$
$$\zeta^{l_{ia}m_{ia}}_{l_{ib}m_{ib}} = \int d\vec{r}_1 \int d\vec{r}_2 \frac{\zeta^{l_{ia}}_{m_{ia}}\left(\vec{r}_1 - \vec{a}\right)\zeta^{l_{ib}}_{m_{ib}}\left(\vec{r}_2 - \vec{b}\right)}{\left|\vec{r}_1 - \vec{r}_2\right|} \tag{30}$$

By explicitly considering angular momentum states, these orbitals naturally integrate the rotational symmetry inherent in molecular systems, leading to optimized quantum representations and computational scalability [50, 51]. This angular momentum basis also reveals a deep and nontrivial connection between orbital coupling, quantum entanglement, and algorithmic complexity. In particular, entanglement between atomic centers arises through the coupling of their angular momenta, which becomes especially complicated under long-range Coulomb interactions.

Unlike an isolated atom, where all relevant spatial coordinates and field points can be conveniently defined in a unified spherical frame of reference, a polyatomic molecular system often lacks clear global symmetry. Consequently, each atomic center defines its own local frame, and the relative orientation between these local frames must be explicitly considered in molecular integrals. This dependence on geometric relationships between multiple centers significantly enhances the entanglement structure of the molecular wavefunction. Specifically, the relative orientation of axes connecting distinct atomic centers governs the strength and complexity of angular momentum coupling. Such couplings inherently generate substantial quantum entanglement between different orbital angular momentum states. Under Coulombic fields, this entanglement becomes even more pronounced due to multipole expansions of the electron-electron interaction. Each multipole term induces additional angular momentum coupling, linking orbital angular momenta across different centers with the angular momentum states associated with the Coulomb field itself. Thus, the presence of long-range Coulomb interactions intensifies the entanglement network, making it increasingly challenging for classical and current quantum computers to accurately represent and manipulate these highly entangled states due to exponential computational resource demands.

This entanglement, however, precisely underpins the potential advantage of quantum computing for quantum chemistry simulations. Quantum computers naturally encode and process entangled states without incurring exponential memory overhead, suggesting they can efficiently manage complex angular momentum couplings inherent in molecular systems. The critical question, therefore, is why and how quantum computers leverage entanglement for computational acceleration. Entanglement is not simply a computational obstacle; rather, it serves as a fundamental quantum resource enabling quantum algorithms to achieve significant speedups compared to their classical counterparts. Thus, developing angular momentum-aware algorithms that explicitly exploit and efficiently manipulate this entanglement will be essential in fully harnessing the power of quantum computers for molecular simulations.

The electrostatic landscape of any molecular or solid-state system is ultimately governed by solutions to the Poisson equation (for unscreened 1/r interactions) and its screened counterpart, the Helmholtz equation (relevant to dielectric media and periodic boundary conditions). Accurate, scalable solutions to these partial differential equations are indispensable for evaluating Coulomb and exchange integrals, constructing effective Hamiltonians, and capturing long-range correlation effects. We have developed an efficient algorithm for solving the potential of the Coulomb integrand $\zeta\left(\vec{r}-\vec{a}\right)$ in Eq. (27), which satisfies the Poisson equation,

$$\nabla^2\varphi_a\left(\vec{r}-\vec{a}\right)=-4\pi\zeta\left(\vec{r}-\vec{a}\right) \tag{31}$$

In this algorithm, we first applied solid harmonics to decompose $\zeta\left(\vec{r}-\vec{a}\right)$into multipole components,

$$\zeta\left(\left|\vec{r}-\vec{a}\right|\right)_{lm}=\frac{1}{\left|\vec{r}-\vec{R}_a\right|^{l+1}}\int d\Omega Y^l_m\left(\vec{r}-\vec{R}_a\right)\zeta\left(\vec{r}-\vec{a}\right) \tag{32}$$

where the integral element $d\Omega$ is on the solid angle. The corresponding potentials $\varphi_{lm}(r)$ of multipole components with projected densities $\zeta\left(\left|\vec{r}-\vec{a}\right|\right)_{lm}$ can be evaluated through the Green's function,

$$\varphi_{lm}(r)=\begin{cases}\frac{1}{r^{2l+1}}\int_0^r dxx^{l+2}\zeta(x)_{lm}+\int_r^{R_a^{(0)}} dxx^{2-(l+1)}\zeta(x)_{lm}, r<R_a^{(0)}\ \frac{q_{lm}}{r^{2l+1}}, r>R_a^{(0)}\end{cases} \tag{33}$$

where $R_a^{(0)}$the radius of atomic spheres and q_{lm} the multipole moment charge. With the cubic spline interpolation method, these finite integrals $\int_a^b dxx^n$ can be analytically evaluated such as $\frac{1}{n+1}\left(b^{n+1}-a^{n+1}\right)$ where n is an integer. Finally, the Coulomb potential $\varphi_a\left(\vec{r}-\vec{a}\right)$ corresponding to partitioned electronic densities $\xi\left(\vec{r}-\vec{a}\right)$ are

$$\varphi_a\left(\vec{r}-\vec{a}\right)=\sum\nolimits_{lm}Y^l_m\left(\vec{r}-\vec{a}\right)\varphi_{lm}\left(\left|\vec{r}-\vec{a}\right|\right)$$

This algorithm is also applicable to Helmholtz equations,

$$\nabla^2\varphi\left(\vec{r}\right)\pm\lambda^2\varphi\left(\vec{r}\right)=-4\pi\rho\left(\vec{r}\right)$$

where the integrands in Eq. (27) are either spherical modified Bessel functions or Hankel and Bessel functions, depending on the modified Helmholtz equation or normal Helmholtz equation. In terms of quantum effects related to light entanglement and quantum information technology, the normal Helmholtz equation is necessary for light-matter interactions. For extended large systems, the algorithm scales as $O(N)$ because the potential $\varphi_a\left(\vec{r} - \vec{a}\right)$ in $r > R_a^{(0)}$ can be evaluated by using the Ewald sum technique. Therefore, the quantum algorithms developed in this paper have broader applications using quantum computers.

Lastly, a solid harmonic product, such as that employed in Eq. (20), provides a structured means of characterizing the entanglement arising from electron-electron interactions under Coulomb fields. This entanglement structure can be naturally expressed using spherical tensors, which have found successful application in modern machine learning architectures, particularly spherical convolutional neural networks (CNNs) [52, 53]. These spherical CNNs leverage spherical harmonics, a complete orthonormal basis on a sphere, which inherently form the irreducible representations of the 3D rotation group SO [3]. By construction, spherical CNNs exhibit rotational equivariance, enabling highly efficient and accurate modeling of rotationally symmetric molecular properties such as atomization energies, electronic densities, and molecular wavefunctions.

The potential advantage of spherical CNNs in molecular modeling not only relies on rotational equivariance but is also fundamentally linked to their ability to implicitly manage orbital angular momentum entanglement. Specifically, spherical harmonic expansions impose constraints on the highest angular momentum contributions in Coulombic fields, thereby simplifying complex entanglement patterns inherent in multi-center molecular systems. For instance, CG neural networks, a specialized class of spherical CNNs, explicitly encode angular momentum coupling rules through hierarchical CG transformations. In predicting molecular wavefunctions with deep CG networks, angular momentum channels typically extend up to 2 l to adequately represent molecular Hamiltonians constructed from atomic orbitals with maximum angular momentum l. This upper-bound angular momentum requirement aligns precisely with the fundamental physical limit set by orbital angular momentum entanglement induced by Coulomb fields within molecules, underscoring the deep relationship between machine learning frameworks and physical quantum chemistry.

Moreover, the mathematical structure underpinning CG networks mirrors the angular momentum coupling and uncoupling encountered in quantum many-body problems. Specifically, CG networks implement linear transformations of CG products for their Fourier vector outputs, effectively decoupling angular momentum channels to simplify the representation of highly entangled molecular states. Thus, the interplay between solid harmonic expansions, CG neural networks and quantum algorithm design forms a coherent and powerful strategy. This synergy can facilitate the construction of robust quantum and classical computational frameworks that capture complex molecular phenomena efficiently and accurately, establishing a firm theoretical and practical foundation for next-generation molecular simulations.

5. Conclusions

This chapter introduces a novel computational framework leveraging Solid Harmonic Gaussian Orbitals (SHGOs) to significantly enhance the efficiency and accuracy of molecular integral evaluations. Unlike traditional Cartesian Gaussian Orbitals (CGOs), SHGOs explicitly incorporate angular momentum symmetries, clearly separating radial and angular dependencies, which allows for more efficient computational strategies. By employing advanced vector-coupling and vector-uncoupling schemes, along with Clebsch-Gordan coefficients and Wigner 3-j symbols, our approach systematically reduces the computational complexity typically associated with molecular integrals. The practical advantages of SHGO-based methods are evident through marked reductions in floating-point operations, positioning them as superior alternatives to conventional computational chemistry techniques.

Beyond computational savings, the SHGO methodology aligns seamlessly with quantum computing paradigms. Angular momentum entanglement networks constructed through Clebsch-Gordan transformations provide robust platforms for quantum simulations, opening pathways to efficiently tackle previously intractable molecular systems. This alignment not only enhances computational scalability and accuracy in traditional quantum chemistry applications but also extends into emerging fields such as quantum sensing and imaging. By explicitly handling angular momentum, SHGO-based frameworks bridge classical computational methods with state-of-the-art quantum computational technologies, promising significant advancements across both theoretical and practical domains.

The presented SHGO approach thus represents a critical advancement in computational quantum chemistry, effectively addressing computational limitations, enhancing predictive accuracy, and providing a versatile foundation for future scientific and technological explorations.

Acknowledgements

The authors gratefully acknowledge the financial support provided by Defense Research and Development Canada (DRDC) and the National Research Council (NRC) of Canada. This work was made possible through funding from NRC's Applied Quantum Challenge program, which enabled the exploration and development of advanced computational methods at the intersection of quantum chemistry and quantum information science.

Conflict of interest

The authors declare no conflict of interest.

Appendices and nomenclature

KS-DFT	Kohn-Sham density functional theory
SHGOs	solid harmonic Gaussian orbitals
CGOs	Cartesian Gaussian orbitals
CG	Clebsch-Gordan
CNNs	convolutional neural networks
THC	tensor hypercontraction
VQE	variational quantum Eigensolver
QPE	quantum phase estimation

Author details

Hang Hu[1,2], Gilles Peslherbe[1*], Hsu Kiang Ooi[2] and Anguang Hu[3*]

1 Centre for Research in Molecular Modeling and Department of Chemistry and Biochemistry, Concordia University, Montreal, Quebec, Canada

2 Data Science for Complex Systems, National Research Council of Canada, Toronto, ON, Canada

3 Suffield Research Center, Defence Research Development Canada, Medicine Hat, AB, Canada

*Address all correspondence to: gilles.peslherbe@concordia.ca; anguang.hu@drdc-rddc.gc.ca

References

[1] Clementi E. Methods and Techniques in Computational Chemistry: METECC. 94th ed. B: Medium Size System. STEF; 1993. 574 p

[2] McArdle S, Endo S, Aspuru-Guzik A, Benjamin SC, Yuan X. Quantum computational chemistry. Reviews of Modern Physics. 2020;**92**(1):015003

[3] Mardirossian N, Head-Gordon M. Thirty years of density functional theory in computational chemistry: An overview and extensive assessment of 200 density functionals. Molecular Physics. 2017;**115**(19):2315-2372

[4] Condon EU, Shortley G. The Theory of Atomic Spectra. Cambridge: Cambridge University Press; 1935

[5] Slater JC. The theory of complex spectra. Physics Review. 1929;**34**(10): 1293-1322

[6] Hartree DR. The wave mechanics of an atom with a non-coulomb central field. Part I. Theory and methods. Mathematical Proceedings of the Cambridge Philosophical Society. 1928; **24**(1):89-110

[7] Kohn W, Becke AD, Parr RG. Density functional theory of electronic structure. The Journal of Physical Chemistry. 1996; **100**(31):12974-12980

[8] Hohenberg P, Kohn W. Inhomogeneous electron gas. Physics Review. 1964;**136**(3B):B864-B871

[9] Kohn W, Sham LJ. Self-consistent equations including exchange and correlation effects. Physical Review. 1965;**140**(4A):A1133-A1138

[10] Zabusky NJ. Solitons and bound states of the time-independent Schrodinger equation. Physics Review. 1968;**168**(1):124-128

[11] Zahradník R, Polák R. The time-independent Schrödinger equation. In: Zahradník R, Polák R, editors. Elements of Quantum Chemistry [Internet]. Boston, MA: Springer US; 1980. pp. 13-46. DOI: 10.1007/978-1-4613-3921-2_3

[12] Schrödinger E. An undulatory theory of the mechanics of atoms and molecules. Physics Review. 1926;**28**(6): 1049-1070

[13] Bohr N. Über die Anwendung der Quantentheorie auf den Atombau. Zeitschrift für Physik. 1923;**13**(1):117-165

[14] Bohr N. The quantum postulate and the recent development of atomic Theory1. Nature. 1928;**121**(3050):580-590

[15] Atkins P, de Paula J, Friedman R. Quanta, Matter, and Change: A Molecular Approach to Physical Chemistry. Oxford: OUP Oxford; 2009

[16] Mccaw CS. Orbitals: With Applications in Atomic Spectra. Singapore: World Scientific Publishing Company; 2015

[17] Sakurai JJ, Commins ED. Modern quantum mechanics, revised edition. American Journal of Physics. 1995;**63**(1): 93-95

[18] Edmonds AR. Angular Momentum in Quantum Mechanics. Princeton: Princeton University Press; 1996

[19] Avery JE, Avery JS. Hyperspherical Harmonics and their Physical Applications [Internet]. Singapore: World Scientific; 2017. 300 p. DOI: 10.1142/10690

[20] Hollas JM. Modern Spectroscopy. Etobicoke, Ontario, Canada: John Wiley & Sons; 2004

[21] Gill PMW. Molecular integrals over Gaussian basis functions. In: Sabin JR, Zerner MC, editors. Advances in Quantum Chemistry [Internet]. Pittsburgh: Academic Press; 1994. pp. 141-205. Available from: https://www.sciencedirect.com/science/article/pii/S0065327608600192

[22] Boys SF, Egerton AC. Electronic wave functions - I. A general method of calculation for the stationary states of any molecular system. Proceedings of the Royal Society of London Series A Mathematical and Physical Sciences. 1997;**200**(1063):542-554

[23] Schlegel HB, Frisch MJ. Transformation between Cartesian and pure spherical harmonic Gaussians. International Journal of Quantum Chemistry. 1995;**54**(2):83-87

[24] Pople JA, Hehre WJ. Computation of electron repulsion integrals involving contracted Gaussian basis functions. Journal of Computational Physics. 1978; **27**(2):161-168

[25] McMurchie LE, Davidson ER. One- and two-electron integrals over Cartesian Gaussian functions. Journal of Computational Physics. 1978;**26**(2): 218-231

[26] Dunlap BI. Three-center Gaussian-type-orbital integral evaluation using solid spherical harmonics. Physical Review A. 1990;**42**(3):1127-1137

[27] Dunlap BI. Angular momentum in molecular quantum mechanical integral evaluation. Computer Physics Communications. 2005;**165**(1):18-36

[28] Hu A, Dunlap BI. Three-center molecular integrals and derivatives using solid harmonic Gaussian orbital and Kohn–Sham potential basis sets. Canadian Journal of Chemistry. 2013; **91**(9):907-915

[29] Dunlap BI, Zope RR. Efficient quantum-chemical geometry optimization and the structure of large icosahedral fullerenes. Chemical Physics Letters. 2006;**422**(4–6):451-454

[30] Matveev AV, Nasluzov VA, Rösch N. Linear response formalism for the Douglas–Kroll–Hess approach to the Dirac–Kohn–Sham problem: First- and second-order nuclear displacement derivatives of the energy. International Journal of Quantum Chemistry. 2007; **107**(15):3236-3249

[31] Nikodem A, Matveev AV, Soini TM, Rösch N. Load balancing by work–stealing in quantum chemistry calculations: Application to hybrid density functional methods. International Journal of Quantum Chemistry. 2014;**114**(12):813-822

[32] Hobson EW. The Theory of Spherical and Ellipsoidal Harmonics. Cambridge: University Press; 1955

[33] Hu A, Staufer M, Birkenheuer U, Igoshine V, Rösch N. Analytical evaluation of pseudopotential matrix elements with Gaussian-type solid harmonics of arbitrary angular momentum. International Journal of Quantum Chemistry. 2000;**79**(4): 209-221

[34] Boys SF, Cook GB, Reeves CM, Shavitt I. Automatic fundamental calculations of molecular structure. Nature. 1956;**178**(4544):1207-1209

[35] Arfken GB, Weber HJ, Harris FE. Bessel functions. In: Arfken GB, Weber HJ, Harris FE, editors. Mathematical Methods for Physicists

(Seventh Edition) [Internet]. Boston: Academic Press; 2013. pp. 643-713. Available from: https://www.sciencedirect.com/science/article/pii/B9780123846549000141

[36] Murphy KV, Turney JM, Henry F, III S. Student-friendly guide to molecular integrals. Journal of Chemical Education [Internet]. 2018;**95**:1572-1578. DOI: 10.1021/acs.jchemed.8b00255

[37] Obara S, Saika A. Efficient recursive computation of molecular integrals over Cartesian Gaussian functions. The Journal of Chemical Physics. 1986;**84**(7): 3963-3974

[38] Babbush R, Wiebe N, McClean J, McClain J, Neven H, Chan GKL. Low-depth quantum simulation of materials. Physical Review X. 2018;**8**(1):011044

[39] Cao Y, Romero J, Olson JP, Degroote M, Johnson PD, Kieferová M, et al. Quantum chemistry in the age of quantum computing. Chemical Reviews. 2019;**119**(19):10856-10915

[40] Peruzzo A, McClean J, Shadbolt P, Yung MH, Zhou XQ, Love PJ, et al. A variational eigenvalue solver on a photonic quantum processor. Nature Communications. 2014;**5**(1):4213

[41] Bharti K, Cervera-Lierta A, Kyaw TH, Haug T, Alperin-Lea S, Anand A, et al. Noisy intermediate-scale quantum algorithms. Reviews of Modern Physics. 2022;**94**(1):015004

[42] Tilly J, Chen H, Cao S, Picozzi D, Setia K, Li Y, et al. The variational quantum Eigensolver: A review of methods and best practices. Physics Reports. 2022;**986**:1-128

[43] Parrish RM, Hohenstein EG, McMahon PL, Martínez TJ. Quantum computation of electronic transitions using a variational quantum Eigensolver. Physical Review Letters. 2019;**122**(23): 230401

[44] Abrams DS, Lloyd S. Quantum algorithm providing exponential speed increase for finding eigenvalues and eigenvectors. Physical Review Letters. 1999;**83**(24):5162-5165

[45] Kang C, Bauman NP, Krishnamoorthy S, Kowalski K. Optimized quantum phase estimation for simulating electronic states in various energy regimes. Journal of Chemical Theory and Computation. 2022;**18**(11): 6567-6576

[46] Yamamoto K, Duffield S, Kikuchi Y, Muñoz RD. Demonstrating Bayesian quantum phase estimation with quantum error detection. Physical Review Research. 2024;**6**(1):013221

[47] Parker JB, Joseph I. Quantum phase estimation for a class of generalized eigenvalue problems. Physical Review A. 2020;**102**(2):022422

[48] Veis L, Višňák J, Nishizawa H, Nakai H, Pittner J. Quantum chemistry beyond born–Oppenheimer approximation on a quantum computer: A simulated phase estimation study. International Journal of Quantum Chemistry. 2016;**116**(18):1328-1336

[49] Lee J, Berry DW, Gidney C, Huggins WJ, McClean JR, Wiebe N, et al. Even more efficient quantum computations of chemistry through tensor hypercontraction. PRX Quantum. 2021;**2**(3):030305

[50] Helgaker T, Jorgensen P, Olsen J. Molecular Electronic-Structure Theory. Wiley; 2008

[51] Shavitt I, Bartlett RJ. Many-Body Methods in Chemistry and Physics:

MBPT and Coupled-Cluster Theory [Internet]. Cambridge: Cambridge University Press; 2009. Available from: https://www.cambridge.org/core/books/manybody-methods-in-chemistry-and-physics/D12027E4DAF75CE8214671D842C6B80C

[52] Kondor R, Lin Z, Trivedi S. Clebsch-Gordan Nets: A Fully Fourier Space Spherical Convolutional Neural Network [Internet]. 2018. Available from: http://arxiv.org/abs/1806.09231

[53] Cohen TS, Geiger M, Koehler J, Welling M. Spherical CNNs [Internet]. 2018. Available from: http://arxiv.org/abs/1801.10130

Section 3

Quantum Hardware

Chapter 5

Superconducting Parametric Amplifiers: Resonator Design and Role in Qubit Readout

Babak Mohammadian

Abstract

Superconducting parametric amplifiers (SPAs) are critical components for ultra-low-noise qubit readout in quantum computing, addressing the critical challenge of amplifying weak quantum signals without introducing noise that degrades coherence and computational fidelity. Unlike classical amplifiers, SPAs can achieve or closely approach quantum-limited performance, specifically the *Standard Quantum Limit* (SQL) of half a photon of added noise for phase-preserving amplification. The core principle of SPAs relies on parametric amplification, where energy is transferred from a strong pump tone to a weak input signal through non-dissipative nonlinear mixing processes. This is enabled by intrinsic nonlinearities in superconducting materials, primarily kinetic inductance in thin films (e.g., NbTiN, Al) and, more significantly, the Josephson effect in Josephson junctions. These nonlinear elements facilitate frequency mixing (three-wave or four-wave mixing) and can operate in phase-preserving or phase-sensitive amplification modes, with the latter allowing for noise squeezing below the SQL. This chapter emphasizes the significant role of resonator design in determining critical SPA performance metrics such as gain, bandwidth, and noise characteristics. It details both lumped-element (LC) and distributed-element (coplanar waveguide, CPW) resonators, discussing their unique properties, suitability for different frequency ranges, and the importance of achieving high-quality factors (Q) for efficient energy storage and minimal loss. A practical design and simulation of a meandered quarter-wavelength CPW resonator coupled to a feed line is presented, illustrating how precise control over geometric parameters optimizes resonant frequency, coupling strength, and quality factor for high-fidelity qubit state discrimination.

Keywords: superconducting parametric amplifiers, Josephson junctions, quantum-limited readout, coplanar waveguide resonators, nonlinear mixing, full-wave electromagnetic simulation

1. Introduction

Reading out quantum states with high precision while minimizing added noise remains one of the foremost challenges in quantum computing. Qubits are inherently fragile and can be easily perturbed by unwanted interactions, particularly during measurement. This noise degrades coherence and reduces the computational fidelity.

Therefore, ultra-low-noise amplification is essential to preserve quantum information and improve readout accuracy [1, 2].

Superconductivity provides ultra-low-noise quantum measurements. Defined by zero electrical resistance and the Meissner effect [3], superconductors exhibit extremely low dissipation at cryogenic temperatures. For quantum circuits, operating at millikelvin temperatures is critical—not simply to achieve superconductivity, but to suppress thermal excitations and maintain quantum coherence. These conditions enable high-quality factors (Q) in microwave resonators used for qubit control and readout [4, 5]. Modern superconducting processors employ thin-film fabrication to realize complex circuits containing Josephson junctions and high-Q resonators [6], which provide low-loss energy storage and nonlinear interactions essential for quantum operations [7, 8].

In superconducting processors, qubit signals are extremely weak—on the order of only a few photons at microwave frequencies (4–8 GHz). Detecting these signals without excessive added noise is crucial for high-fidelity state discrimination. Consequently, the first amplification stage must provide high gain, minimal added noise, and ideally, quantum-limited performance [9, 10]. Resonator design within parametric amplifiers strongly influences these metrics, as geometry and material properties dictate bandwidth, gain flatness, and noise.

Traditional amplifiers such as High Electron Mobility Transistors (HEMTs) [11], even at cryogenic temperatures, exhibit noise temperatures in the 1–5 K range [12, 13]. As classical semiconductor devices, HEMTs are constrained by thermal and transistor noise mechanisms and cannot approach quantum-limited performance. For quantum systems, this added noise is prohibitive: Weak microwave signals corresponding to a few photons are easily buried under several kelvins of amplifier noise. By contrast, quantum-limited amplification requires an added noise of only about half a photon at the signal frequency—the Standard Quantum Limit (SQL)—the minimum noise permitted by quantum mechanics for a phase-preserving amplifier [14, 15]. Noise temperature here quantifies the equivalent thermal noise power introduced by the amplifier, providing an intuitive comparison of noise performance.

Superconducting parametric amplifiers (SPAs) can closely approach the quantum limit, making them indispensable for quantum measurement. They function by transferring energy from a strong pump tone to a weak input signal through nonlinear mixing. This process enables amplification with minimal added noise, particularly when using superconducting materials such as NbTiN and Al to utilize nonlinearities arising from kinetic inductance or the Josephson effect [16, 17]. The nonlinear inductance of Josephson junctions is especially critical, as it directly enables parametric amplification. In superconducting qubit platforms, SPAs are integrated into the readout chain, where the qubit state is dispersively encoded in a resonator signal that must be amplified and distinguished from noise [18, 19]. Resonator design in SPAs therefore plays a central role in setting performance metrics, including the achievable signal-to-noise ratio (SNR).

Moreover, parametric amplifiers enable quantum-limited phase-sensitive detection [20] and noise squeezing [21], which can suppress noise in one quadrature below the SQL. This capability is particularly valuable in quantum error correction [22] and quantum sensing applications [23], where maximizing information extraction while minimizing disturbance is essential.

This chapter focuses on the design and operation of superconducting parametric amplifiers, with emphasis on resonator-based architectures. We examine how resonator geometry and material properties determine key performance metrics such as gain, bandwidth, and noise. We also discuss the role of SPAs in dispersive qubit readout chains and their impact on the achievable signal-to-noise ratio (SNR). By combining theoretical foundations with practical design considerations, the chapter provides a comprehensive perspective on how resonator design enables near-quantum-limited amplification in superconducting processors.

2. Superconducting parametric amplification

A parametric amplifier consists of three fundamental components. The first is a nonlinear element, which is essential for approaching the Standard Quantum Limit (SQL). To achieve near-quantum-limited performance, the amplifier must employ a non-dissipative nonlinearity that enables frequency mixing without adding excess noise. In contrast to linear systems, where signals propagate without modifying one another, nonlinear systems allow interactions that generate new frequency components and complex dynamics [17]. Such nonlinearities can arise from current-dependent inductances [24], voltage-dependent capacitances, or quantum effects in Josephson junctions [25, 26].

The second component is the pump: a strong microwave or RF tone that supplies the external energy driving the parametric process [10]. Finally, a resonant circuit is typically employed to enhance the interaction between the signal and the nonlinear element, thereby improving gain and frequency selectivity by confining the operation to well-defined modes [27].

2.1 Superconductivity and kinetic inductance

Superconductivity arises when a material is cooled below its critical temperature T_c, a state characterized by zero electrical resistance and expulsion of magnetic fields —the Meissner effect [28]. Within BCS theory, this phase emerges from Cooper pairs, electron pairs bound *via* phonon-mediated attraction, which condense into a macroscopic quantum state [29]. Below T_c, an energy gap $\Delta(T)$ opens at the Fermi surface, suppressing scattering and enabling dissipationless current flow.

The electromagnetic response of superconductors is described by the London equations [30]:

$$\frac{\partial \vec{J}_s}{\partial t} = \frac{1}{\mu_0 \lambda^2} \vec{E}, \qquad \nabla \times \vec{J}_s = -\frac{1}{\lambda^2} \vec{H}, \tag{1}$$

where λ is the London penetration depth. These equations capture the essential features of perfect conductivity and magnetic field expulsion.

Despite having zero DC resistance, superconductors exhibit a finite *kinetic inductance* at nonzero frequencies, arising from the inertia of Cooper pairs [31]. When a time-varying current flows, Cooper pairs must be accelerated, requiring energy and producing a reactive impedance that stores energy in carrier motion rather than magnetic fields. For small currents, this inductance remains nearly constant; at higher currents, however, pair-breaking induces nonlinear behavior:

$$L_K(I) = L_K(0)\left(1 + \frac{I^2}{I_*^2} + \cdots\right), \tag{2}$$

where I_* is a characteristic current scale [24, 32]. This intrinsic nonlinearity makes kinetic inductance a valuable mechanism for superconducting parametric amplification.

2.2 Josephson inductance

Josephson inductance provides another nonlinear mechanism employed in parametric amplification [33]. It originates from the Josephson junction, which consists of two superconducting electrodes separated by a nanometer-scale insulating barrier. Although separated by this barrier, Cooper pairs can tunnel quantum mechanically from one electrode to the other without resistance [25], as illustrated in **Figure 1**.

The supercurrent through the junction follows the first Josephson relation:

$$I = I_c \sin \phi, \tag{3}$$

where I_c is the critical current and ϕ is the superconducting phase difference. This nonlinear current-phase relation underpins the use of Josephson junctions in nonlinear circuits and parametric amplifiers.

The second Josephson relation links the time evolution of the phase to the voltage across the junction:

$$\frac{d\phi}{dt} = \frac{2e}{\hbar} V \quad \text{or equivalently} \quad V = \frac{\Phi_0}{2\pi} \frac{d\phi}{dt}, \tag{4}$$

where $\Phi_0 = h/(2e)$ is the magnetic flux quantum. By combining the two Josephson relations and comparing with the definition of an inductor, $V = L_J dI/dt$, one obtains:

$$L_J = \frac{\Phi_0}{2\pi I_c \cos \phi}. \tag{5}$$

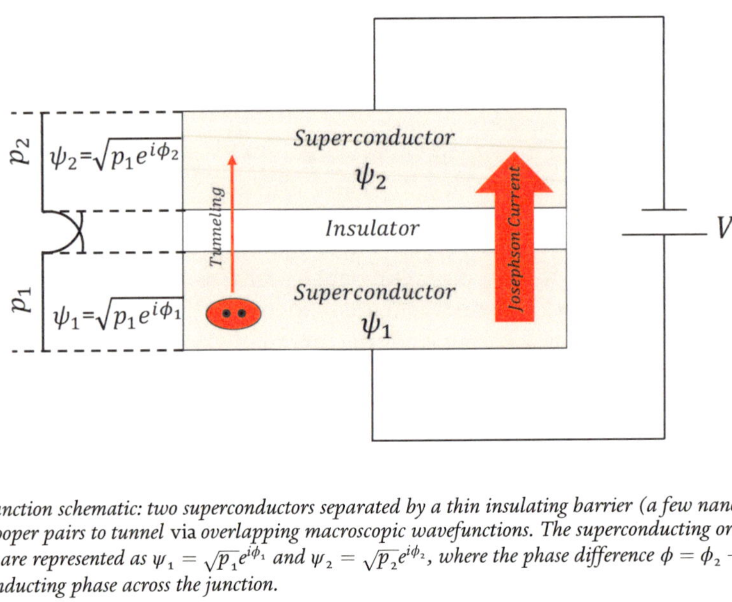

Figure 1.
Josephson junction schematic: two superconductors separated by a thin insulating barrier (a few nanometers), allowing Cooper pairs to tunnel via *overlapping macroscopic wavefunctions. The superconducting order parameters are represented as $\psi_1 = \sqrt{p_1} e^{i\phi_1}$ and $\psi_2 = \sqrt{p_2} e^{i\phi_2}$, where the phase difference $\phi = \phi_2 - \phi_1$ defines the superconducting phase across the junction.*

This expression shows that the Josephson inductance L_J is both nonlinear and tunable, since it depends explicitly on the phase difference ϕ across the junction. As ϕ approaches $\pi/2$, $\cos\phi$ vanishes and the inductance diverges, producing strong nonlinearity. Moreover, because L_J is inversely proportional to the critical current I_c, its value can be engineered by adjusting junction parameters or controlled dynamically using an external flux bias (**Table 1**).

Feature	Kinetic inductance	Josephson inductance
Physical origin	Inertia of Cooper pairs	Quantum tunneling of Cooper pairs
Nonlinearity type	Kerr-type (third order)	Depends on circuit design and pumping
Key parameter	Current-dependent suppression of gap	Phase difference across junction ϕ
Mathematical form	$L_k(I)\approx L_{k0}(1+\alpha I^2)$	$L_J = \frac{\Phi_0}{2\pi I_c \cos\phi}$
Tuning mechanism	Fixed by material properties	Tunable *via* magnetic flux in SQUIDs
Typical implementation	Superconducting transmission lines	Josephson junctions/SQUIDs
Applications	SPAs, kinetic inductance detectors	SPAs, qubits, tunable resonators

Note: Φ_0 is the magnetic flux quantum; I_c is the junction critical current. The effective nonlinearity type of Josephson inductance depends on the circuit configuration and pumping scheme. It can be engineered for second-order nonlinearity (e.g., in flux-pumped SQUIDs for 3WM) or third-order (Kerr-type) nonlinearity (e.g., in current-pumped single junctions for 4WM).

Table 1.
Comparison of kinetic and Josephson inductance nonlinearities.

2.3 Quantum noise and the standard quantum limit

Quantum amplifiers are fundamentally constrained by the uncertainty principle, which sets a lower bound on the simultaneous precision with which two conjugate quadratures of an electromagnetic mode can be defined. For a single mode with canonical quadratures $\hat{X}$ and $\hat{P}$, the uncertainty relation is:

$$\Delta X\,\Delta P \geq \frac{\hbar}{2}. \tag{6}$$

Here, ΔX and ΔP denote the standard deviations of the quadrature operators. This relation implies that any phase-preserving linear amplifier—which amplifies both quadratures equally—must inevitably add noise in order to satisfy the uncertainty bound [14].

The *Standard Quantum Limit* (SQL) defines this minimum added noise, corresponding to the equivalent of half a photon at the signal frequency. This sets the ultimate performance bound for any phase-preserving amplifier permitted by quantum mechanics. In terms of noise temperature, the SQL is:

$$T_{\text{SQL}} = \frac{\hbar\omega}{2k_{\text{B}}}, \tag{7}$$

where ω is the angular frequency of the signal and k_{B} is the Boltzmann's constant. For qubit readout frequencies in the 4–8 GHz range, this corresponds to a noise temperature on the order of 100 mK—roughly an order of magnitude lower than that of the best cryogenic HEMTs (1–5 K) [12, 34].

2.4 Phase-preserving vs. phase-sensitive amplification

Quantum amplifiers are generally classified into two categories: phase-preserving and phase-sensitive [14].

- *Phase-preserving amplifiers* amplify both quadratures of the input signal equally. To satisfy the uncertainty principle, they must add noise, with the SQL setting a minimum of half a photon of added noise power.

- *Phase-sensitive amplifiers* amplify one quadrature while attenuating the other. In this case, amplification can occur without adding noise to the amplified quadrature, enabling below-SQL performance in that quadrature at the expense of increased uncertainty in the conjugate one. This mechanism produces quadrature noise squeezing, which is valuable in quantum metrology and quantum error correction [35].

2.5 Gain and added noise

One key performance metric for an amplifier is its *power gain* G, defined as

$$G = \frac{P_{\text{out}}}{P_{\text{in}}}, \tag{8}$$

where P_{out} and P_{in} denote the output and input signal powers, respectively.

High gain alone is insufficient for quantum applications; the added noise must also be minimized. For phase-preserving amplifiers, the fundamental link between gain and noise is captured by the Haus-Caves limit [14], which states:

$$N_{\text{add}} \geq \frac{1}{2}\hbar\omega(G-1), \tag{9}$$

where N_{add} denotes the added noise referred to the input, expressed as an equivalent energy at the signal frequency [14]. This inequality shows that although higher gain improves the signal-to-noise ratio (SNR) in subsequent stages, it cannot eliminate the intrinsic quantum noise penalty of phase-preserving amplification.

3. Superconducting parametric amplifiers

To approach the SQL, amplifiers must employ a non-dissipative nonlinear element to enable frequency mixing without introducing significant noise.

Parametric amplification is the process by which the amplitude of a weak signal is increased by transferring energy from a strong external pump through a nonlinear medium. Unlike conventional amplifiers that rely on active semiconductor elements such as transistors, parametric amplifiers use the time-dependent variation of a system parameter—such as inductance or capacitance—to mediate energy exchange [14].

The primary figure of merit in parametric amplification is the *gain*, which quantifies the increase in signal power. Parametric amplifiers can typically achieve high gain (20–30 dB) with minimal added noise. However, this performance comes at the expense of *bandwidth*, which is approximately inversely proportional to the gain due to the fundamental gain-bandwidth tradeoff [36].

For resonator-based parametric amplifiers, the gain-bandwidth tradeoff can be expressed as:

$$\sqrt{G}\Delta f \approx \kappa, \tag{10}$$

where G is the power gain, Δf is the amplifier bandwidth, and κ is the resonator linewidth (its energy decay rate, in Hz). Thus, as the gain increases, the bandwidth narrows so that the product remains approximately constant.

3.1 Frequency mixing

When a strong pump tone at frequency ω_p and a weak signal at frequency ω_s are applied to a nonlinear medium—such as a superconducting resonator with kinetic or Josephson inductance—new frequency components, known as *idler* tones (ω_i), are generated according to

$$\omega_i = \omega_p \pm \omega_s. \tag{11}$$

This process, called frequency mixing, originates from the time-dependent modulation of the system's reactive impedance and is a direct consequence of nonlinearity [37, 38].

In nonlinear mixing, the idler frequency enables efficient energy transfer from the pump to the signal. The pump provides energy that is shared between the signal and idler, ensuring energy conservation and maintaining phase coherence. This allows signal amplification without introducing extra noise from dissipative processes. Without the idler, the pump could not effectively transfer energy, and amplification would be largely ineffective [35, 38].

To better understand how nonlinear systems generate and manipulate new frequency components, it is useful to distinguish between two fundamental mechanisms —three-wave mixing and four-wave mixing—illustrated in **Figure 2**. These processes arise from different orders of nonlinearity and obey distinct energy-conservation relations [39, 40].

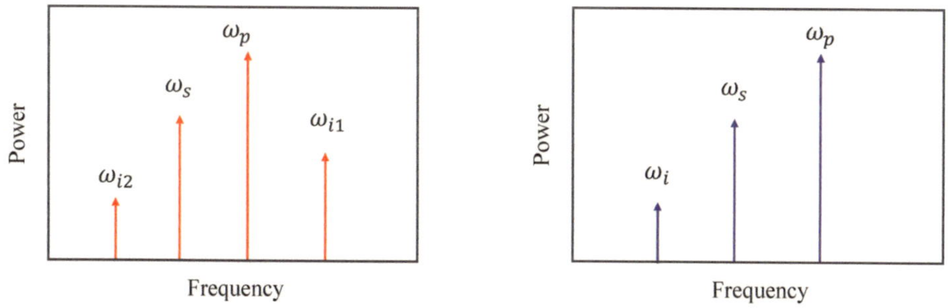

Figure 2.
Three-wave and four-wave mixing. The pump (ω_p) and signal (ω_s) tones interact within a nonlinear resonant medium. (Left) In four-wave mixing, third-order nonlinearity generates two idler frequencies, $\omega_{i1} = 2\omega_p - \omega_s$ and $\omega_{i2} = 2\omega_s - \omega_p$, leading to frequency conversion and amplification. (Right) In three-wave mixing, second-order nonlinearity produces a single idler at $\omega_i = \omega_p - \omega_s$, also enabling frequency conversion and amplification. The idler power depends on both gain and pump strength.

3.2 Degenerate vs. non-degenerate parametric amplification

As discussed above, parametric amplification is a nonlinear process in which energy from a pump tone is transferred to a signal and an idler. Depending on the frequency relationship between these tones, two main regimes arise: *degenerate*, where the signal and idler frequencies coincide, and *nondegenerate*, where they remain distinct [41, 42].

Degenerate parametric amplification occurs when the signal and idler frequencies coincide, that is, $\omega_s = \omega_i = \frac{\omega_p}{2}$. This requires both tones to lie within the same resonator mode, with the pump frequency approximately twice the resonator's natural frequency. Because the signal and idler share the same mode, the amplifier operates in a *phase-sensitive* regime, where the gain depends on the relative phase between the input signal and the pump. In this case, one quadrature of the signal can be amplified, while the orthogonal quadrature (90° out of phase) is attenuated. This enables noise squeezing below the Standard Quantum Limit (SQL), which is particularly valuable for quantum-limited measurements [43].

In contrast, *nondegenerate parametric amplification* occurs when the signal and idler occupy different modes of the circuit. These frequencies often couple through separate physical ports, each associated with its own resonance. In this configuration, the amplifier can provide (i) *reflection gain*, amplifying signals reflected at the same frequency, and (ii) *conversion gain*, transferring energy between the signal and idler frequencies [36, 43]. Unlike the degenerate case, nondegenerate amplification is largely phase-insensitive: It amplifies both quadratures equally and cannot achieve quadrature-selective gain or squeezing. Consequently, such amplifiers add at least half a photon of noise per quadrature [43].

3.3 Types of parametric amplifiers

In superconducting parametric amplifiers, signal amplification is achieved through wave-mixing processes driven by the device's intrinsic nonlinearity. This nonlinearity may arise from Josephson inductance (reactive) or from kinetic inductance, which can also involve dissipative contributions [44]. Accordingly, superconducting parametric amplifiers are broadly categorized into two main classes based on the underlying nonlinearity: Josephson Parametric Amplifiers (JPAs) [17, 45] and Kinetic Inductance Parametric Amplifiers (KIPAs) [38]. Both types can operate in either four-wave mixing (4WM) or three-wave mixing (3WM) regimes, depending on the form of nonlinearity used in the device.

3.3.1 Josephson Parametric Amplifiers (JPAs)

Josephson Parametric Amplifiers (JPAs) employ the intrinsic nonlinear inductance of Josephson junctions to achieve low-noise amplification of microwave signals. The effective inductance of a Josephson junction depends on the current flowing through it, allowing it to act as a nonlinear circuit element. The core principle is parametric amplification: A strong pump tone drives the junction into a nonlinear regime, periodically modulating the resonator's effective inductance. This modulation transfers energy from the pump to the signal, enabling amplification with minimal added noise (**Figure 3**) [45, 46].

JPAs operate through frequency mixing driven by a strong pump tone, with distinct regimes depending on circuit design and nonlinearity.

In the *three-wave mixing (degenerate) mode*, a pump at $\omega_p \approx 2\omega_s$ drives the circuit, producing signal (ω_s) and idler (ω_i) tones that overlap at the same frequency

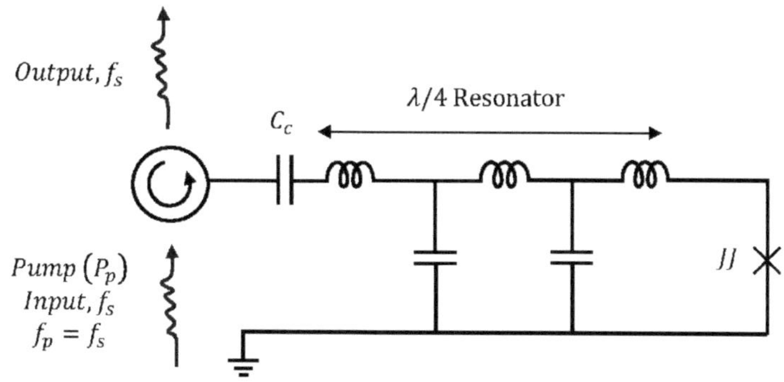

Figure 3.
Lumped-element schematic representation of a resonant Josephson Parametric Amplifier (JPA). The device operates in reflection mode, with both the input signal (f_s) and pump tone (P_p) applied to the same port. A coupling capacitor (C_c) connects the input/output line to a $\lambda/4$ resonator modeled as a lumped LC network. At the grounded end, a Josephson junction (JJ) provides nonlinear inductance, enabling parametric amplification. In the degenerate three-wave configuration, the pump frequency is approximately twice the signal frequency ($f_p \approx 2f_s$). In practice, a circulator or isolator is placed at the port to separate the reflected amplified signal from the incoming tone.

($\omega_s = \omega_i$). This regime enables *phase-sensitive amplification*, where only one quadrature of the microwave field is amplified, ideally without introducing added noise. It is typically realized in *flux-pumped JPAs*, in which a Superconducting Quantum Interference Device (SQUID) terminates a resonator and its inductance is modulated by a flux drive at $2\omega_s$ [37, 47].

By contrast, in the *four-wave mixing (non-degenerate) mode*, the pump frequency is close to the signal frequency ($\omega_p \approx \omega_s$). Energy conservation requires $2\omega_p = \omega_s + \omega_i$, so the signal and idler emerge at distinct frequencies ($\omega_s \neq \omega_i$). This regime provides *phase-preserving amplification*, in which both quadratures are amplified equally. Such operation is commonly implemented in *current-pumped JPAs*, where a pump tone directly drives a single Josephson junction through a coupling capacitor [36, 43]. However, phase-preserving amplification is unavoidably constrained by the *Standard Quantum Limit*, introducing at least half a photon of added noise [14].

3.3.2 Traveling waveguide parametric amplifiers

While JPAs provide near-quantum-limited amplification within a narrow bandwidth, their resonant nature inherently limits their frequency range and dynamic range [48]. To overcome these constraints, *Traveling Wave Parametric Amplifiers* (TWPAs) have been developed as a broadband alternative [49].

The operation of TWPAs is fundamentally based on the principle of parametric excitation, where energy is transferred to an oscillatory system by periodically varying one of its parameters, such as capacitance or inductance. For example, in a simple resonant circuit, if the capacitance is modulated in time—such as pulling apart condenser plates when the voltage reaches a maximum and restoring them when the voltage is zero—energy is added to the circuit, sustaining oscillations. This parametric modulation is typically driven at twice the resonant frequency of the circuit and can be implemented as a square wave or sinusoidal variation. In TWPAs, this concept is extended to a nonlinear transmission line, where the nonlinearity of Josephson junctions effectively modulates the system parameters dynamically as the pump tone propagates, enabling continuous energy transfer to the signal (**Figure 4**).

Unlike JPAs, which rely on a resonant structure, TWPAs use a nonlinear transmission line composed of a series of Josephson junctions or nonlinear elements that allow a pump tone to propagate along the line and continuously interact with the signal based on the principle of four-wave mixing to satisfy the frequency relation $2\omega_p = \omega_s + \omega_i$. This distributed nonlinearity facilitates parametric amplification over a wide frequency range without the bandwidth restrictions imposed by resonators [10, 38].

The continuous nature of the nonlinear medium allows the amplification to build up gradually over the entire length of the device, enabling high gain (often exceeding 20 dB) and large bandwidths (several GHz) simultaneously [50].

TWPAs typically employ arrays of Josephson junctions embedded in coplanar waveguides or lumped-element transmission lines, with careful impedance engineering to maintain low insertion loss and high dynamic range. Their broadband, low-noise performance has made them indispensable for high-fidelity readout of superconducting qubits and other quantum devices [10, 36].

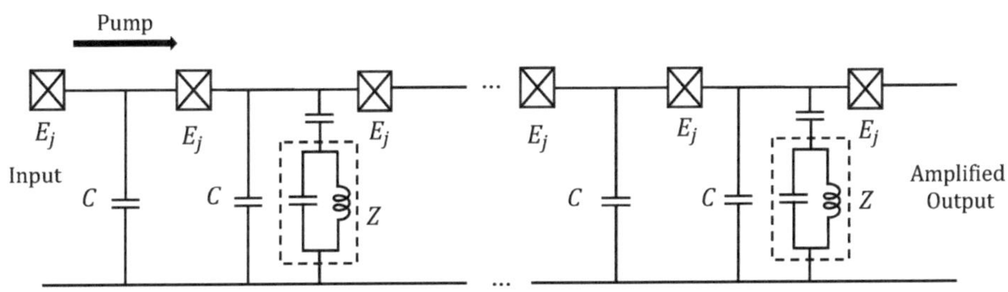

Figure 4.
Schematic representation of a Traveling-Wave Parametric Amplifier (TWPA) based on a Josephson junction transmission line. The amplifier consists of a series of Josephson junctions (cross symbols) and capacitive shunt elements, with periodically inserted phase-matching structures (inductor-capacitor networks). The input signal enters on the left, is amplified through parametric interaction with a strong pump tone, and exits as an amplified output on the right.

3.3.3 Kinetic Inductance Parametric Amplifiers

Kinetic Inductance Parametric Amplifiers (KIPAs) are the key components in quantum information processing, especially for high-sensitivity quantum measurements. They are typically deployed as the first amplification stage, operating at the lowest temperature to provide sufficient gain and overcome noise from subsequent stages, such as high-electron-mobility-transistor (HEMT) amplifiers [51].

KIPAs operate based on the nonlinear kinetic inductance of thin superconducting films (e.g., NbTiN or NbN), which arises from the inertia of Cooper pairs. This nonlinearity becomes more pronounced in thin films and under current bias. Under an applied current, the kinetic inductance L_k becomes current-dependent and is often approximated as:

$$L_k(I) \approx L_{k0}\left(1 + \alpha I^2\right), \quad (12)$$

where L_{k0} is the zero-bias kinetic inductance and α characterizes the nonlinearity.

In these devices, four-wave mixing (4WM) can occur in transmission lines patterned as coplanar waveguides or microstrips when driven by a strong pump tone. Phase matching is required to maximize gain across the desired bandwidth, which is typically achieved through dispersion engineering. By tailoring the geometric dispersion, phase matching is maintained while suppressing unwanted higher-order processes, such as the pump's third harmonic.

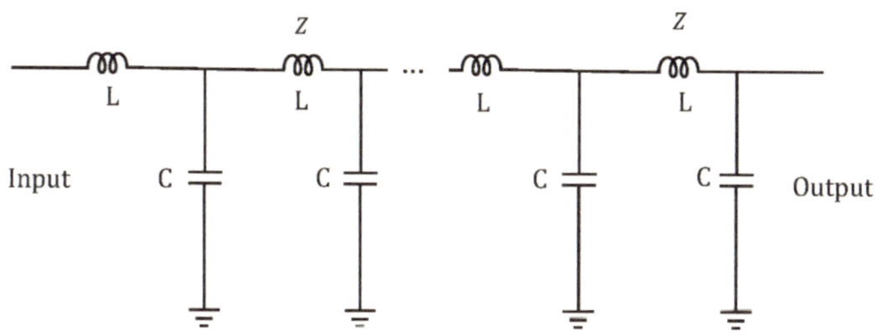

Figure 5.
Layout of a Kinetic Inductance Parametric Amplifier (KIPA) based on a nonlinear transmission line. The structure consists of alternating coplanar waveguide (CPW) sections with engineered impedance and embedded kinetic inductance elements. The Z segments are cascaded to form a stepped-impedance profile, enabling phase matching and parametric gain. These segments can be designed as $\lambda/4$ or $\lambda/2$ CPWs.

Applying a DC bias allows three-wave mixing (3WM) in kinetic inductance traveling-wave parametric amplifiers (KI-TWPAs), yielding near-quantum-limited noise performance, broad bandwidth, and high dynamic range (**Figure 5**) [52, 53].

Having explored the physical principles and architectural variations of parametric amplifiers—including JPAs, TWPAs, and KIPAs—it is essential to understand how

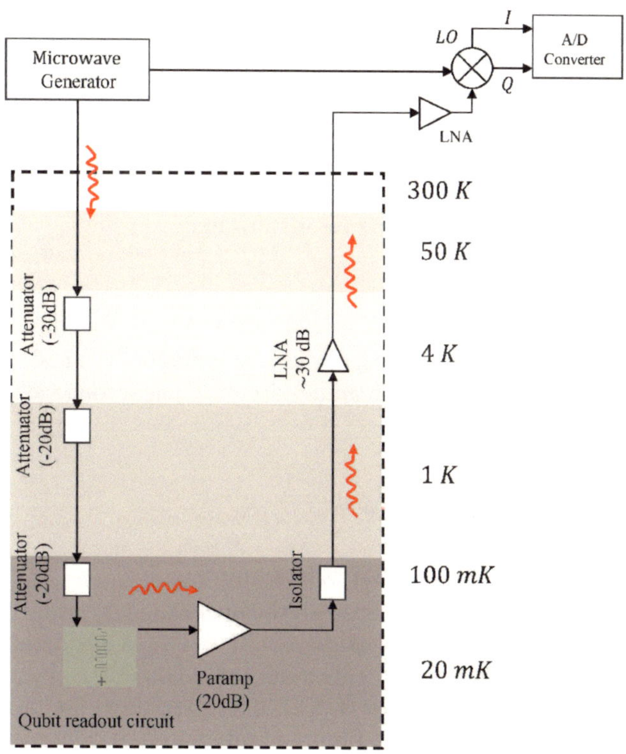

Figure 6.
Cryogenic measurement setup for superconducting qubit readout: A probe signal from a vector network analyzer (VNA) passes through a series of attenuators at progressively colder temperature stages to suppress thermal noise. At the 20 mK stage, the signal interacts with the qubit circuit and is amplified by a parametric amplifier (Paramp) with ~ 20 dB gain. An isolator at 100 mK prevents back-action noise from higher stages. The signal is further amplified by a low-noise amplifier (LNA) at 4 K and then demodulated via an I/Q mixer driven by a local oscillator (LO). The final signal is digitized by an analog-to-digital converter (A/D), enabling high-fidelity qubit state discrimination.

these devices are integrated into a cryogenic measurement chain for qubit readout. The following schematic illustrates a typical experimental setup used in superconducting quantum circuits, where signal integrity, noise suppression, and thermal isolation are critical. Each amplifier type discussed earlier plays a distinct role depending on its placement, gain characteristics, and compatibility with the surrounding microwave infrastructure (**Figure 6**).

4. Superconducting resonators: Structure and function in parametric amplifiers

4.1 Basics of microwave resonators

Microwave resonators are fundamental building blocks in superconducting parametric amplifiers (SPAs), providing energy storage and defining the interaction between microwave signals and nonlinear elements. A resonator confines electromagnetic energy and oscillates at discrete frequencies where constructive interference occurs—reflected waves combine in phase to reinforce each other [54]. In SPAs, the resonator enhances the pump tone and facilitates energy transfer between the pump, signal, and idler modes [38].

Superconducting resonators operate by forming standing waves at specific resonant frequencies. The use of superconducting materials is essential, as their unique properties, particularly the superconducting gap, ensure extremely low energy dissipation, yielding very high-quality factors. To minimize external disturbances such as thermal noise and stray radiation, these devices are typically operated at millikelvin temperatures inside dilution refrigerators [55, 56].

The main types of superconducting resonators are:

4.1.1 Lumped-element (LC) resonators

Lumped-element resonators consist of discrete inductors (L) and capacitors (C) connected in series or parallel to form an LC circuit. Their resonance frequency is given by:

$$f_0 = \frac{1}{2\pi\sqrt{LC}}. \tag{13}$$

These resonators provide a compact solution for low-frequency applications, where the physical size is much smaller than the operating wavelength [57]. At microwave frequencies, however, parasitic elements such as lead inductance, inter-turn capacitance, and conductor losses (including the skin effect) become significant and limit the achievable quality factor (Q). To overcome these limitations, modern designs employ superconducting thin films and integrated parallel-plate capacitors, reducing parasitics and achieving Q factors exceeding 10^5 while occupying an area of approximately 1 mm^2 [58].

4.1.2 Distributed-element (CPW) resonators

Distributed-element resonators, such as coplanar waveguide (CPW) resonators, are implemented using sections of transmission line designed to support standing-wave boundary conditions, typically $\lambda/4$ or $\lambda/2$ [59]. Unlike lumped-element designs, the inductance and capacitance in CPW resonators are distributed along the length of the line, eliminating the need for discrete components. The characteristic impedance Z_0 (commonly around 50 Ω) and propagation properties are determined by the inductance and capacitance per unit length of the transmission line [60].

While a CPW resonator's behavior at resonance can often be approximated by a lumped LC circuit for simplified analysis, its inherent distributed nature fundamentally governs all its resonant modes and frequency-dependent characteristics, both on- and off-resonance. CPW resonators are widely used in superconducting quantum circuits due to their ease of design up to frequencies of approximately 10 GHz, immunity to stray parasitics, and excellent compatibility with planar fabrication processes. When implemented using superconducting films at cryogenic temperatures, these resonators can achieve loaded quality factors ranging from a few hundred to several hundred thousand [59].

In CPW resonators, the electric field is concentrated in the gaps between the center conductor and the ground planes, with the magnetic field primarily surrounding the center conductor. This geometry allows precise control of impedance and coupling by adjusting the gap and conductor width. Engineering the field profile is essential for optimizing nonlinear participation in SPAs (**Figure 7**) [61].

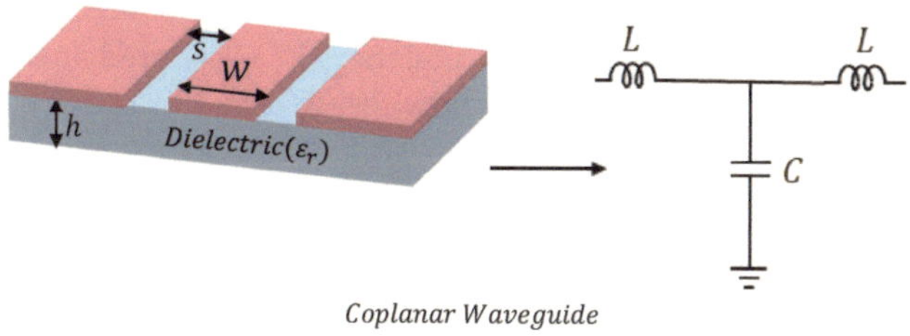

Figure 7.
Schematic of a coplanar waveguide (CPW) resonator and its equivalent lumped-element model. The CPW consists of a central conductor of width W separated by gaps s from the ground planes on either side, all superconducting material patterned on a dielectric substrate of thickness h and relative permittivity ε_r.

4.1.3 $\lambda/4$ and $\lambda/2$ resonators

Quarter-wave ($\lambda/4$) and half-wave ($\lambda/2$) configurations refer to the resonator length relative to the guided wavelength of the electromagnetic wave it supports. The guided wavelength is given by:

$$\lambda = \frac{v_p}{f} = \frac{c}{f\sqrt{\varepsilon_{\text{eff}}}}, \tag{14}$$

where v_p is the phase velocity of the wave in the transmission line, and ε_{eff} is the effective dielectric constant of the coplanar waveguide (CPW) structure [54].

A $\lambda/2$ resonator, often realized as a half-wavelength section of transmission line, supports a standing wave with voltage nodes (and current antinodes) at both ends for the fundamental mode. Its fundamental resonance occurs when the physical length satisfies:

$$l \approx \frac{\lambda}{2}, \quad f_0 = \frac{v_p}{2l}. \tag{15}$$

In contrast, a $\lambda/4$ resonator is a quarter-wavelength section, typically with one end short-circuited to ground and the other end open or capacitively coupled to the feed line. In this configuration, the shorted end forms a voltage node, while the open end forms a voltage antinode for the fundamental mode, and the resonance condition is:

$$l \approx \frac{\lambda}{4}, \quad f_0 = \frac{v_p}{4l}. \tag{16}$$

Quarter-wave resonators are preferred when easy coupling to feed lines is desired, while half-wave resonators are chosen for symmetric boundary conditions or multi-mode operation [54, 59].

4.2 Quality factor (Q)

The *quality factor* (Q) quantifies how efficiently a resonator stores electromagnetic energy relative to its losses and characterizes the sharpness of its resonance. It also influences gain, bandwidth, and noise performance [54]. It is defined as

$$Q = 2\pi \frac{\text{Energy Stored}}{\text{Energy Lost per Cycle}} = \frac{\omega_0 W}{P_{\text{loss}}}, \tag{17}$$

where $\omega_0 = 2\pi f_0$ is the resonant angular frequency, W the stored energy, and P_{loss} the dissipated power per cycle. In terms of bandwidth Δf:

$$Q = \frac{f_0}{\Delta f}. \tag{18}$$

In superconducting CPW resonators, different loss mechanisms are quantified by distinct quality factors. The internal quality factor, Q_{int}, accounts for intrinsic losses such as conductor residual resistance, dielectric absorption, and radiation, and is defined by $Q_{\text{int}} = \frac{\omega_0 W}{P_{\text{internal}}}$. The external quality factor, Q_{ext}, describes losses due to intentional coupling to external circuitry, such as feed lines, and is given by $Q_{\text{ext}} = \frac{\omega_0 W}{P_{\text{external}}}$.

When both internal and external losses are present, the total or loaded quality factor, Q_L, satisfies $\frac{1}{Q_L} = \frac{1}{Q_{\text{int}}} + \frac{1}{Q_{\text{ext}}}$.

The loaded Q determines the operational bandwidth and energy decay rate of the resonator.

5. Design and simulation of a CPW resonator-feed line system for qubit readout

In this section, we present the design and simulation of a planar superconducting readout circuit based on coplanar waveguide (CPW) resonators coupled to a common feed line.

This architecture, widely used in circuit quantum electrodynamics (cQED), enables efficient dispersive measurement of superconducting qubits in the microwave regime. We focus on a single-qubit configuration in which the qubit is coupled to a $\lambda/4$

meandered CPW resonator, and the resonator is interfaced with a CPW feed line for multiplexed readout. We outline the functional role of each circuit element, discuss key design considerations, and highlight how geometry sets parameters such as resonant frequency, coupling strength, and quality factors (see **Figure 8**).

Figure 8.
A qubit circuit using CPW as a resonator coupled with the feed line modeled in HFSS.

5.1 Design overview

The primary application of this structure is the dispersive readout of superconducting qubits in the microwave domain, particularly for frequency multiplexing. In this scheme, multiple resonators of slightly different lengths are coupled to the same feed line, enabling each qubit to be read out at a distinct frequency.

The following is what each component's role is:

1. *Feed line:*

   The feed line is implemented as a coplanar waveguide (CPW), whose role is to deliver and collect microwave signals from the resonators. It transmits readout pulses to the resonators, collects the reflected or transmitted signals that encode qubit states, and also supports multiplexed readout when multiple resonators are coupled along its length.

   The first step in designing the coupling circuit is to realize a CPW feed line with a characteristic impedance of 50 Ω, ensuring proper matching to measurement equipment and minimizing reflections. The substrate is silicon with relative permittivity $\varepsilon_r = 11.7$ and thickness $h = 550, \mu$m. The CPW impedance is set by the conductor width w and gap s, which are iteratively adjusted to achieve the desired impedance. The resulting optimized dimensions are illustrated in **Figure 7**.

   Geometric definitions:

   Here, w is the width of the center conductor, s the gap between the center conductor and the ground planes, h the substrate thickness, and ε_r the relative permittivity of the substrate.

   We define the normalized geometric ratios:

$$k_0 = \frac{w}{w+2s}, \qquad k_1 = \frac{\sinh\left(\frac{\pi w}{4h}\right)}{\sinh\left(\frac{\pi(w+2s)}{4h}\right)} \tag{19}$$

Elliptic integral approximations:

The complete elliptic integrals of the first kind, $K(k_0)$ and $K(k_1)$, are approximated using logarithmic expressions depending on the value of k [62]:
For $0 \leq k \leq 0.71$*:*

$$K(k) \approx \left[\frac{1}{\pi}\ln\left(\frac{2\left(1+\sqrt{\sqrt{1-k^2}}\right)}{1-\sqrt{\sqrt{1-k^2}}}\right)\right]^{-1} \tag{20}$$

For $0.71 < k \leq 1$*:*

$$K(k) \approx \frac{1}{\pi}\ln\left(\frac{2\left(1+\sqrt{k}\right)}{1-\sqrt{k}}\right) \tag{21}$$

These approximations allow us to compute the ratio $K(k_1)/K(k_0)$, which governs both the impedance and the effective permittivity.

Effective dielectric constant:

The effective dielectric constant is given by:

$$\varepsilon_{\text{eff}} = 1 + \frac{\varepsilon_r - 1}{2} \cdot \frac{K(k_1)}{K(k_0)} \tag{22}$$

This accounts for the field distribution between the substrate and air regions.

Characteristic impedance:

The characteristic impedance of the CPW is:

$$Z_0 = \frac{30\pi}{\sqrt{\varepsilon_{\text{eff}}}} \cdot \frac{1}{K(k_0)} \tag{23}$$

This analytical expression assumes negligible conductor thickness and an ideal lossless dielectric. In practice, it does not capture the effects of finite metallization thickness, field interactions with air in open-back or suspended substrates, or deviations arising from fabrication tolerances and material imperfections. For accurate design, full-wave electromagnetic simulations or calibrated measurements are therefore recommended.
Considering the equations 19 to 21, for or case $w = 10\ \mu\text{m}$ and $s = 6\ \mu\text{m}$, which yield approximately 50 Ω impedance (**Figure 9**).

2.*Meander:*

The meandered resonator functions as a frequency-selective element with a designed resonance in the 4–8 GHz range. It couples weakly to the feed line at one end and to the qubit at the other. In the dispersive regime, the qubit state induces a small shift in the resonator frequency, which can be monitored *via* the feed line to

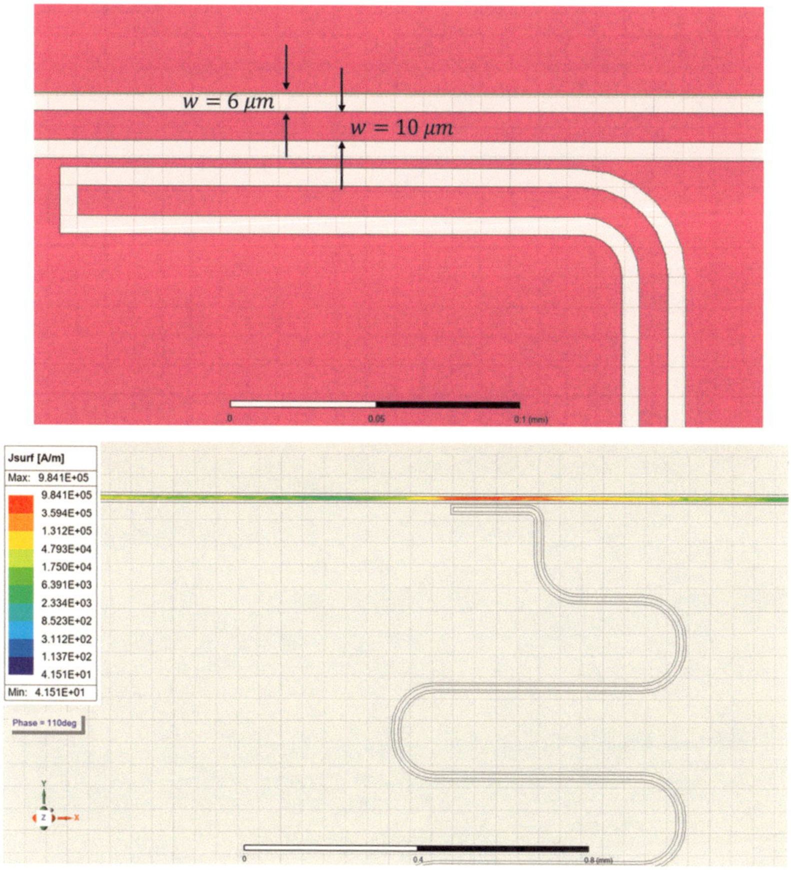

Figure 9.
Feedline: (Top) feed line geometry (Bottom) simulated J-field related to feed line.

infer the qubit state non-destructively. A meandered geometry is employed to increase the effective resonator length while conserving chip area.

The resonator is implemented as a $\lambda/4$ coplanar waveguide (CPW) resonator fabricated from a thin superconducting film, modeled as a perfect conductor in initial simulations. The conductor width w and gap s are chosen to target a characteristic impedance of 50 Ω, with dimensions ($w = 10\ \mu$m, $s = 6\ \mu$m) consistent with those of the feed line. To achieve the desired resonance frequency near 5.5 GHz while remaining compatible with typical chip dimensions, the resonator is realized as a quarter-wavelength transmission line. One end of the line is shorted to ground, while the opposite end is open and capacitively coupled to the transmon qubit. These boundary conditions yield a resonance frequency of

$$f_r \approx \frac{c}{4l\sqrt{\varepsilon_{\text{eff}}}}, \tag{24}$$

where l is the resonator length (in our case ~ 5.43 mm), and $\varepsilon_{\text{eff}} = 6.3$ is the effective dielectric constant of the CPW, calculated from Eq. (23).

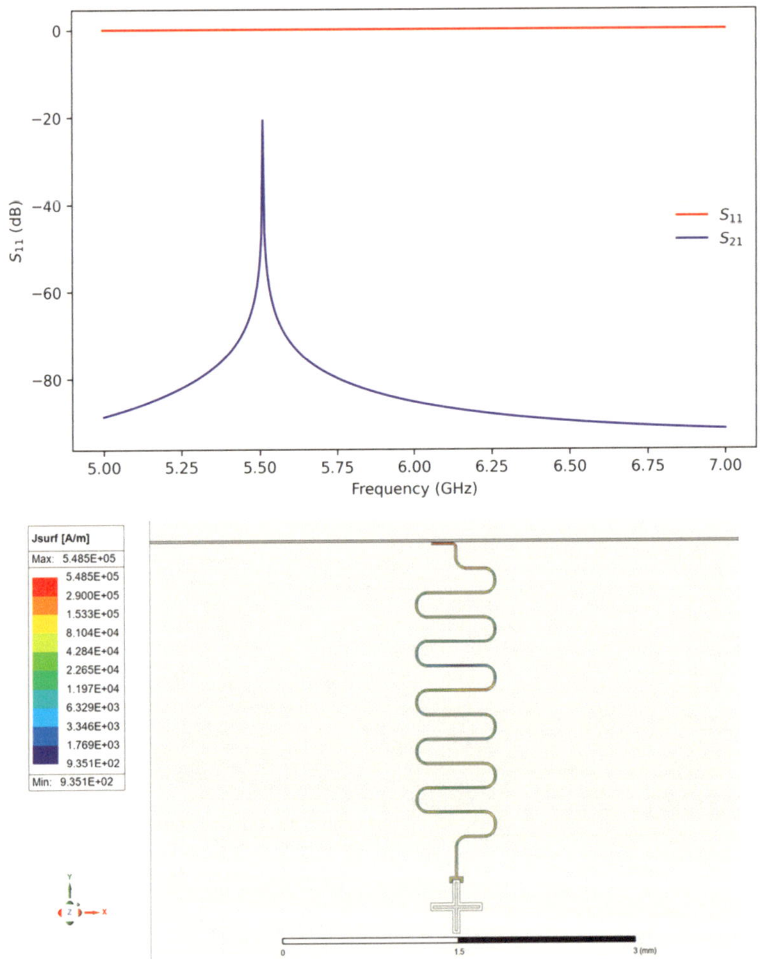

Figure 10.
Meander resonating frequency and related field plot.

3.*Cross-shaped structure:*

The cross-shaped capacitor pads of the transmon qubit provide large shunt capacitance to the Josephson junction, thereby reducing sensitivity to charge noise. They also couple to the resonator, enabling state readout and microwave control. The transmon qubit functions as a nonlinear oscillator, with a transition frequency determined by its Josephson energy (E_J).

Coupling regions:

- *Top short CPW gaps:*

  The feed line-resonator coupling region sets the external quality factor (Q_c) of the resonator. It governs the rate at which energy leaks from the resonator into the feed line, thereby influencing both readout speed and signal strength.

In practice, a CPW resonator is side-coupled to the feed line through a gap capacitor. This capacitance controls the external quality factor Q_c, and thus the coupling strength between the resonator and the measurement line. To calculate the effective capacitance, inductance, and mode frequencies, the CPW geometry (metal widths, spacings, and gaps) is transformed into a simpler equivalent geometry using conformal mapping techniques, where analytical integrals can be evaluated.

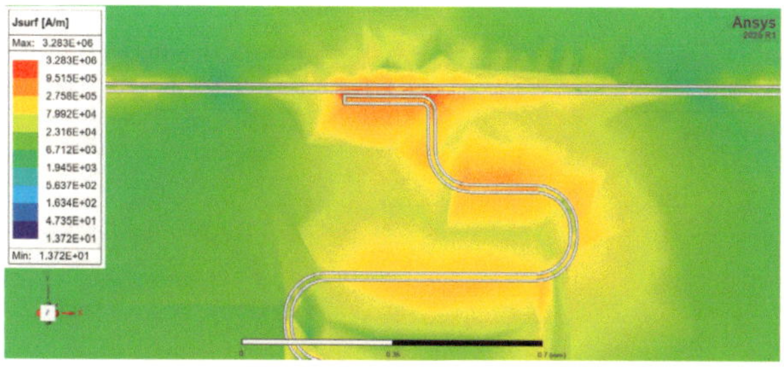

Figure 11.
Field plot of the feed line coupling with meander.

The resonator's electrical length sets the bare quarter-wave resonance frequency, $f_r^{(0)} = \frac{c_\ell}{4(l_o+l_c+l_s)}$, where $c_\ell = c/\sqrt{\varepsilon_{\mathrm{eff}}}$ is the phase velocity on the line and l_o, l_c, and l_s denote the open section, the coupling section, and the shorted section, respectively.

The presence of the feed line introduces an interaction described by a dimensionless coupling coefficient κ, which depends mainly on the coupling gap and overlap, and by two phase factors, $\theta = \frac{2\pi f_r^{(0)} l_c}{c_\ell}$, $\psi = \frac{2\pi f_r^{(0)} (l_c+2l_o)}{c_\ell}$. These phases, together with κ, enter the first-order analytical expressions for the external quality factor Q_e and the frequency shift of the coupled resonator [63].

In our design, considering the geometries shown in **Figures 10–12** and using the analytical description presented in [63], the feed line and resonator impedance are 49.81 and 49.94Ω, $\kappa = 0.093$ with the quality factor of ~ 15000.

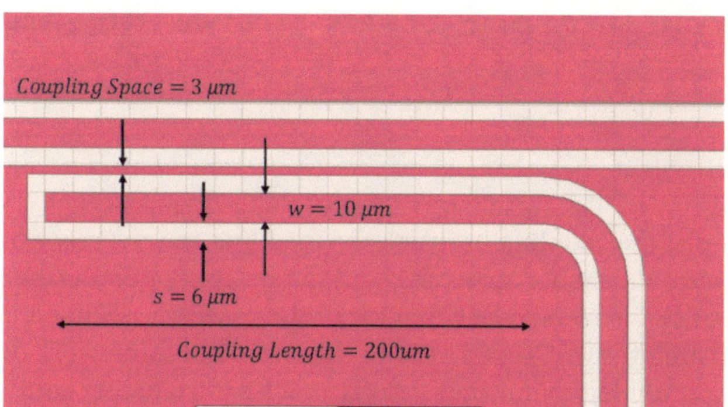

Figure 12.
Meander and feed line coupling geometry.

- *Bottom connection:*

  This region is the resonator-qubit coupling and sets the dispersive coupling rate g, influencing readout contrast and qubit-resonator interaction. To achieve this coupling the same calculation procedure is applied for this part as mentioned above (**Figure 13**).

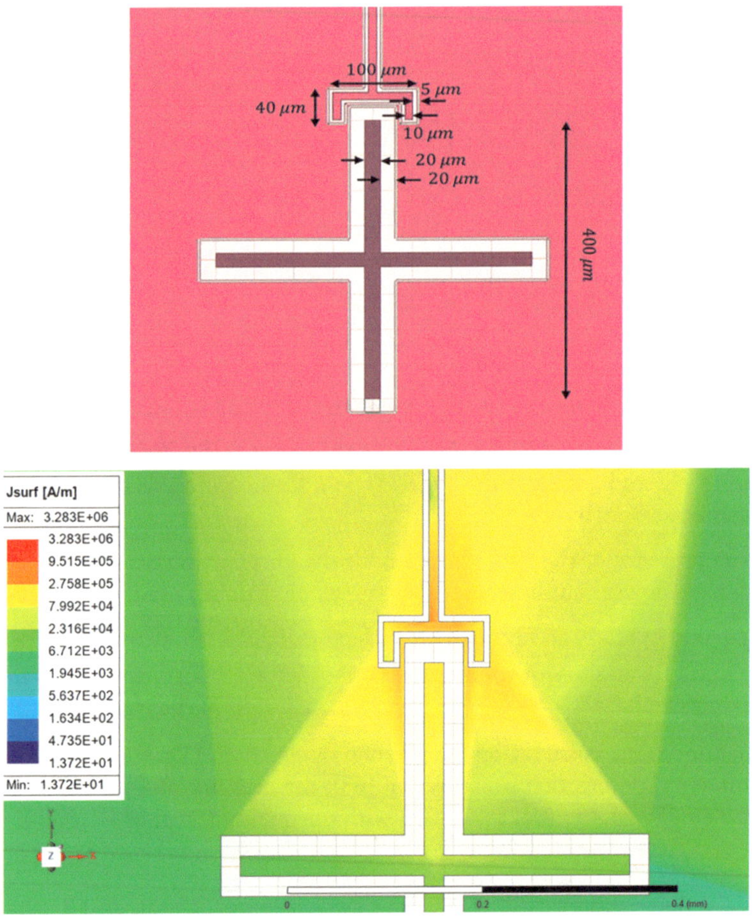

Figure 13.
Cross-shaped capacitor coupling geometry.

In this example, the coupling between the feed line and the resonator was implemented using a meander structure, and the resulting microwave response was analyzed to identify the resonance behavior of the readout, as shown in **Figure 14**.

This readout architecture, together with the analytical framework described above, can be naturally extended to multiple resonators. By introducing additional meandered branches along a single feed line, the system can enable simultaneous readout of several qubits in a scalable and modular fashion, while preserving independent control over coupling strengths and resonance frequencies.

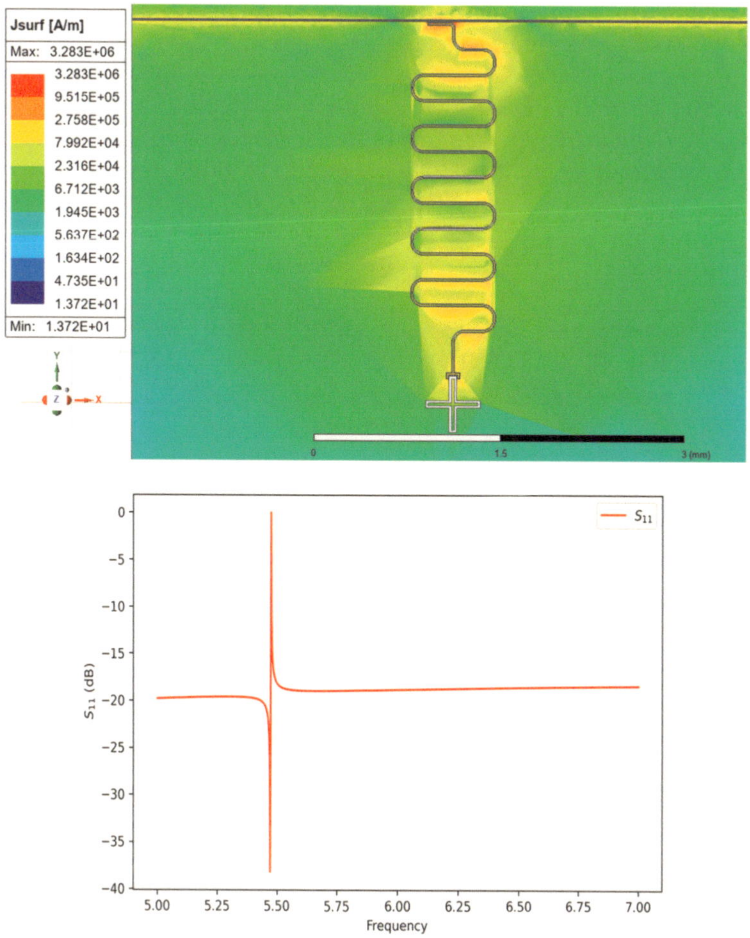

Figure 14.
Design and response of a meander-coupled CPW resonator. (Top) Layout of the coplanar waveguide feed line with a meander section used to control the coupling rate. (Bottom) Simulated S_{11} spectrum showing the resonance frequency, where the depth and linewidth of the dip provide information on the coupling strength and loaded quality factor of the resonator.

6. Conclusion

In this chapter, we have comprehensively explored the foundational principles and practical applications of Superconducting Parametric Amplifiers (SPAs) within quantum computing. We highlighted the critical role of SPAs in achieving ultra-low-noise qubit readout, addressing the limitations of conventional amplifiers and approaching—or even surpassing—the Standard Quantum Limit (SQL).

The chapter reviewed the operating principles of parametric amplification, emphasizing the importance of superconducting nonlinearities from Josephson junctions and kinetic inductance, and distinguishing between phase-preserving and phase-sensitive amplification modes. Resonator design was shown to be central to SPA performance, with both lumped-element and coplanar waveguide structures enabling high-quality factors and optimal energy transfer.

We also examined key SPA architectures, including Josephson Parametric Amplifiers (JPAs), Traveling-Wave Parametric Amplifiers (TWPAs), and Kinetic Inductance Parametric Amplifiers (KIPAs), each offering specific advantages in gain, bandwidth, and dynamic range. Finally, a practical example of a meandered quarter-wavelength CPW resonator coupled to a feed line illustrated the principles of precise engineering for dispersive qubit readout and multiplexed measurement.

Acknowledgements

This work was supported by the Science and Technology Facilities Council (STFC) under grant number ST/X001229/1.

Author details

Babak Mohammadian
Jodrell Bank Centre for Astrophysics, University of Manchester, Manchester, UK

*Address all correspondence to: babak.mohammadian@manchester.ac.uk

References

[1] Nachman B, Urbanek M, de Jong WA, Bauer CW. Unfolding quantum computer readout noise. npj Quantum Information. 2020;**6**:(1):84

[2] Salmanoglu A, Sirat VS. Design of ultra-low noise amplifier for quantum applications (qlna). Quantum Information Processing. 2024;**23**(3):91

[3] Tinkham M. Introduction to Superconductivity. 2nd ed. New York: McGraw-Hill; 1996

[4] Martinis JM. Superconducting phase qubits. Quantum Information Processing. 2009;**8**:81-103

[5] Houck AA, Türeci HE, Koch J. On-chip quantum simulation with superconducting circuits. Nature Physics. 2012;**8**:292-299

[6] Koch J, Yu TM, Gambetta J, Houck AA, Schuster DI, Majer J, et al. Charge-insensitive qubit design derived from the cooper pair box. Physical Review A. 2007;**76**(4):042319

[7] Krantz P, Kjaergaard M, Yan F, Orlando TP, Gustavsson S, Oliver WD. A quantum engineer's guide to superconducting qubits. Applied Physics Reviews. 2019;**6**(2):021318

[8] Clarke J, Wilhelm FK. Superconducting quantum bits. Nature. 2008;**453**(7198):1031-1042

[9] Renger J et al. Quantum-limited josephson traveling-wave parametric amplifier operating at the single-photon level. npj Quantum Information. 2021; 7(1):89

[10] Macklin C, O'Brien K, Hover D, Schwartz ME, Bolkhovsky V, Zhang X, et al. A near-quantum-limited josephson traveling-wave parametric amplifier. Science. 2015;**350**(6258):307-310

[11] Mimura T. The early history of the high electron mobility transistor (hemt). IEEE Transactions on Microwave Theory and Techniques. 2002;**50**(3):780-782

[12] Choi AY, Esho I, Gabritchidze B, Kooi J, Minnich AJ. Characterization of self-heating in cryogenic high electron mobility transistors using schottky thermometry. Journal of Applied Physics. 2021;**130**(15):155107

[13] Mohammadian B, McCulloch MA, Sweetnam T, Gilles V, Piccirillo L. The impact of surface passivation on kapitza resistance at the interface between a semiconductor and liquid nitrogen. Journal of Low Temperature Physics. 2024;**214**(3):125-132

[14] Caves CM. Quantum limits on noise in linear amplifiers. Physical Review D. 1982;**26**(8):1817-1839

[15] Duh KHG, Pospieszalski MW, Kopp WF, Ho P, Jabra AA, Chao P-C, et al. Ultra-low-noise cryogenic high-electron-mobility transistors. IEEE Transactions on Electron Devices. 1988; **35**(3):249-256

[16] McKinstrie CJ, Radic S, Chraplyvy AR. Parametric amplifiers driven by two pump waves. IEEE Journal of Selected Topics in Quantum Electronics. 2002;**8**(3):538-547

[17] Aumentado J. Superconducting parametric amplifiers: The state of the art in josephson parametric amplifiers. IEEE Microwave Magazine. 2020;**21**(8): 45-59

[18] Wendin G. Quantum information processing with superconducting

circuits: A review. Reports on Progress in Physics. 2017;**80**(10):106001

[19] Patel PS, Desai DB. Review of qubit-based quantum sensing. Quantum Information Processing. 2025;**24**(3):1-37

[20] Kakande J. Phase sensitive parametric amplifiers and their applications [PhD thesis]. Southampton, United Kingdom: University of Southampton; 2012

[21] Lee W, Afshari E. Low-noise parametric resonant amplifier. IEEE Transactions on Circuits and Systems I: Regular Papers. 2010;**58**(3):479-492

[22] Lidar DA, Brun TA. Quantum Error Correction. Cambridge, United Kingdom: Cambridge University Press; 2013

[23] Degen CL, Reinhard F, Cappellaro P. Quantum sensing. Reviews of Modern Physics. 2017;**89**(3):035002

[24] Annunziata AJ, Santavicca DF, Frunzio L, Catelani G, Rooks MJ, Frydman A, et al. Tunable superconducting nanoinductors. Nanotechnology. 2010;**21**(44):445202

[25] Josephson BD. Possible new effects in superconductive tunnelling. Physics Letters. 1962;**1**:251-253

[26] Yariv A. Quantum Electronics. 3rd ed. New York, USA: John Wiley & Sons; 1989

[27] Roch N, Flurin E, Nguyen F, Morfin P, Campagne-Ibarcq P, Devoret MH, et al. Widely tunable, nondegenerate three-wave mixing microwave device operating near the quantum limit. Physical Review Letters. 2012;**108**(14):147701

[28] Meissner W, Ochsenfeld R. Ein neuer effekt bei eintritt der supraleitfähigkeit. Naturwissenschaften. 1933;**21**(44):787-788

[29] Bardeen J, Cooper LN, Schrieffer JR. Theory of superconductivity. Physical Review. 1957;**108**(5):1175

[30] London F, London H. The electromagnetic equations of the supraconductor. Proceedings of the Royal Society of London. Series A: Mathematical and Physical Sciences. 1935;**149**(866):71-88

[31] Tinkham M. Introduction to Superconductivity. North Chelmsford, MA, USA: Courier Corporation; 2004

[32] Vissers MR, Hubmayr J, Sandberg M, Chaudhuri S, Bockstiegel C, Gao J. Frequency-tunable superconducting resonators via nonlinear kinetic inductance. Applied Physics Letters. 2015;**107**(6):062601

[33] Kaiser W, Haider M, Russer JA, Russer P, Jirauschek C. Markovian dynamics of josephson parametric amplification. Advances in Radio Science. 2017;**15**:131-136

[34] Krantz P. Phonon black-body radiation limit for heat dissipation in electronics. ResearchGate. 2014. DOI: 10.13140/RG.2.1.1531.8487

[35] Clerk AA, Devoret MH, Girvin SM, Marquardt F, Schoelkopf RJ. Introduction to quantum noise, measurement, and amplification. Reviews of Modern Physics. 2010;**82**(2): 1155-1208

[36] Roy A, Devoret MH. Introduction to parametric amplification of quantum signals with josephson circuits. Comptes Rendus Physique. 2016;**17**(7):740-755

[37] Yamamoto T, Inomata K, Watanabe M, Matsuba K, Miyazaki T,

Oliver WD, et al. Flux-driven josephson parametric amplifier. Applied Physics Letters. 2008;**93**(4):042510

[38] Eom BH, Day PK, LeDuc HG, Zmuidzinas J. Wideband superconducting quantum interference device parametric amplifier. Nature Physics. 2012;**8**(8):623-627

[39] Agrawal GP. Nonlinear Fiber Optics. Boston, MA, USA: Academic Press; 2013

[40] Ernst M, Wulschner F, Lukashenko A, Marx A, Gross R. Kerr nonlinearity in superconducting circuits. Physical Review B. 2014;**90**(6):064508

[41] Parker DJ, Savytskyi M, Vine W, Laucht A, Duty T, Morello A, et al. Degenerate parametric amplification via three-wave mixing using kinetic inductance. Physical Review Applied. 2022;**17**(3):034064

[42] Winkel P, Takmakov I, Rieger D, Planat L, Hasch-Guichard W, Grünhaupt L, et al. Nondegenerate parametric amplifiers based on dispersion-engineered josephson-junction arrays. Physical Review Applied. 2020;**13**(2):024015

[43] Bergeal N et al. Phase-preserving amplification near the quantum limit with a josephson ring modulator. Nature. 2010;**465**(7294):64-68

[44] Zhao S, Withington S, Thomas CN. Superconducting resonator parametric amplifiers with intrinsic separation of pump and signal tones. Journal of Physics D: Applied Physics. 2024;**58**: 35305. Originally posted as arXiv: 2406.02455

[45] Yurke B, Kaminsky PG, Miller RE, Whittaker EA, Smith AD, Silver AH, et al. Observation of 4.2-k equilibrium-noise squeezing via a josephson-parametric amplifier. Physical Review Letters. 1988;**60**(9):764-767

[46] Devoret MH, Schoelkopf RJ. Superconducting circuits for quantum information: An outlook. Science. 2013; **339**(6124):1169-1174

[47] Castellanos-Beltran MA, Lehnert KW. Widely tunable parametric amplifier based on a superconducting quantum interference device array resonator. Applied Physics Letters. 2007; **91**(8):083509

[48] Castellanos-Beltran MA, Irwin KD, Vale LR, Hilton GC, Lehnert KW. Bandwidth and dynamic range of a widely tunable josephson parametric amplifier. IEEE Transactions on Applied Superconductivity. 2009;**19**(3): 944-947

[49] Grebel J, Clerk AA, Cleland AN. Design and performance of a flux-pumped impedance-engineered josephson parametric amplifier. Physical Review Applied. 2021;**16**(3):034046

[50] White TC et al. Traveling wave parametric amplifier with josephson junctions using minimal resonator loading. Applied Physics Letters. 2015; **106**(24):242601

[51] Splitthoff LJ, Wesdorp JJ, Pita-Vidal M, et al. Gate-tunable kinetic inductance parametric amplifier. Physical Review Applied. 2024;**21**(1):014029

[52] Malnou M, Mates JAB, Ranzani L, Vale LR, Hilton GC, Gao J, et al. Three-wave mixing kinetic inductance traveling-wave amplifier with near-quantum-limited noise performance. PRX Quantum. 2021;**2**:010302

[53] Klimovich NS. Traveling wave parametric amplifiers and other nonlinear kinetic inductance devices

[Ph.d. thesis]. Pasadena, CA, USA: California Institute of Technology; 2022

[54] Pozar DM. Microwave Engineering. Hoboken, NJ, USA: Wiley; 2011

[55] Leib M, Deppe F, Marx A, Gross R, Hartmann M. Networks of nonlinear superconducting transmission line resonators. New Journal of Physics. 2012; **14**:075024

[56] Pappas DP, Vissers MR, Wisbey DS, Kline JS, Gao J. Two level system loss in superconducting microwave resonators. In: IEEE Transactions on Applied Superconductivity, Conference Record of the 2010 Applied Superconductivity Conference. Vol. 21. Institute of Electrical and Electronics Engineers (IEEE); 2011. pp. 871-874

[57] Various. Superconducting microwave resonators: Design, fabrication, and applications. arXiv: 2109.07762. 2021

[58] Stockklauser AEA. Design of compact superconducting lc resonators with high quality factor. Physical Review Research. 2023;**5**(4):043126

[59] Göppl M, Fragner A, Baur M, Bianchetti R, Filipp S, Fink JM, et al. Coplanar waveguide resonators for circuit quantum electrodynamics. Journal of Applied Physics. 2008; **104**(11):113904

[60] National Institute of Standards and Technology (NIST). Design and characterization of coplanar waveguide resonators for quantum circuits. In: NIST Technical Note. U.S. Department of Commerce, National Institute of Standards and Technology (NIST); 2015. DOI: 10.6028/NIST.TN.1885

[61] Calusine G, Melville A, Braumüller J, Gudmundsen TJ, Oliver WD. Determining interface dielectric losses in superconducting resonators. Physical Review Applied. 2019;**12**:014012

[62] Wadell BC. Transmission Line Design Handbook. Norwood, MA: Artech House Publishers; 1991

[63] Besedin I, Menushenkov AP. Quality factor of a transmission line coupled coplanar waveguide resonator. EPJ Quantum Technology. 2018;**5**:2